W9-AGC-570

The History of Yachting

The History of Yachting

Douglas Phillips-Birt

Stein and Day
Publishers New York

First published in the United States of
America by Stein and Day/Publishers 1974

This book was designed and produced by
George Rainbird Ltd, Marble Arch House,
44 Edgware Road, London W2

Designers: Harold Bartram and Pat Ariss
Indexer: E. F. Peeler

Text setting, monochrome origination
and printing, and binding
by Cox & Wyman Limited,
Fakenham, Norfolk, England
Colour origination and printing
by Jolly and Barber Limited,
Rugby, Warwickshire, England
Printed in England

ISBN 0–8128–1704–4
Library of Congress Catalog Number 74–76795

Stein and Day/Publishers
Scarborough House, Briarcliff Manor, N.Y. 10510

Frontispiece

The *Sappho* rounding the Vectis, and setting
her balloon canvas, 1870.

Contents

List of color plates

Acknowledgments

The illustrations in this book are designed to amplify rather than decorate the text. The only personal acknowledgment I have to make is to Peter Faure, of George Rainbird Ltd, who collected the illustrations with such energy and perseverance that despite the publisher's generous allowance there had to be some sad discardings at the end. My own collection of yachting illustrations from published and other sources formed a basis on which was built a new assemblage of pictures, some unpublished hitherto, a few rare, others superior reproductions of ones that had appeared before, and all now presented with the good taste of Rainbird's book designers.

We must be grateful to the Royal Ocean Racing Club and the Royal Thames Yacht Club in London for the facilities provided for photographing models or pictures on their premises; also the Royal Yacht Squadron, Cowes, and the Royal Dart Yacht Club, Devon. The New York Yacht Club, some of whose large models had to have their immense glass cases hoisted from them so that their photographs might be taken in as pleasing a way as possible, likewise gave every facility for photography.

I am grateful to publishers and authors for quotations made in the text and acknowledged therein.

A painting by William Van de Velde the Younger of about 1660 showing the yacht *Mary* (left centre) before she was handed over to Charles II but after she had been presented to him. Hence she flies the Union flags and has the English Royal arms on the stern. The other yachts carry the bezan rig as opposed to the spritsail rig of the *Mary*.

Part I Origins to 1850

1 The Dutch way of life

It is usual to offer Cleopatra's Barge, of rather less than 2000 years ago, as an early example of a yacht. This is not unreasonable if too close an inquiry into what is implied by a 'yacht' and 'yachting' is avoided. It also enables a history of yachting to open with the flourish of some of Shakespeare's richest lines.

In the annals of royalty and a few others we run across from time to time down the centuries vessels for which the appropriate word may be 'yacht', in that they were not used for trade, though they may have been for fighting; while they had an elegance whose only function was prestige. Vessels of this character survive in the form of Royal yachts, Presidential yachts, Admiralty yachts, which still belong to various navies, and have been enthusiastically adopted by the navies of the newly emergent nations.

La Blanche Nef was such a vessel, a white craft of fifty oars, and in her occurred an early yachting tragedy. It happened late in the evening of 25 November 1120. The English king, Henry II, was returning from Normandy. He left Barfleur, and *La Blanche Nef*, in which his son Prince William was to sail, prepared to follow. There were on board numerous of the young prince's friends, and also two of Henry's illegitimate children, Richard and May Countess of Perche. It was a gay party, and the gaiety was extended to the crew, who were well treated to wine before sailing.

So, on that quiet, distant November evening the glamour of the English and Norman nobility, twittering sophisticatedly no doubt in the style of A.D. 1120 and humming snatches from the madrigals of the moment, boarded the white yacht and put to sea. The crew pulled strongly hoping to overtake the King's ship. But the vessel suddenly struck a rock and foundered quickly, leaving only one survivor.

That was more than 800 years ago. Why retell the old and famous story? It simply happens to have a peculiar tragic charm – the still evening, the gaiety, the sudden disaster. Most people enjoy a well-constructed tragedy provided it is one of long ago and far away. And the fact that the King is said never to have smiled again is the final, perfect touch. History probably lies in this instance; but with what a perfect sense of style it does so. Yachting has always had a close relationship with style.

The word 'yacht', which has now found its way into so many languages almost without change is, of course, Dutch in origin, and a shortened form of the earlier *jaght schip*. A Dutch–Latin dictionary published in Antwerp in 1599 described a *jaght schip* and a *jaght* (or *joghte*) as a swift light vessel of war, commerce or pleasure. The sporting element in the word lies in the derivation of *jaght* from the root *jaghen*, which meant to hunt, chase or pursue and suggested rapid motion, not necessarily in connection with the sea.

Since an unknown period before 1600 such vessels were used in Holland for purposes largely unrelated to any mode of yachting as later understood. There were State and Admiralty yachts used by government officials and for despatch carrying, and for such work as was later performed by the Revenue Services, the United States Coastguard, the British Trinity House. There were many yachts belonging to the rich Dutch East India Company – one of these was Hudson's famous *D'Halve Maen* (*Half Moon*) which clearly shows the word 'yacht' in its application to a relatively small and handy ship not necessarily employed in sailing for pleasure. The burgomasters owned yachts.

The earliest surviving picture of one is a drawing in Indian ink on parchment by Rool, considerably worn by age; but the faded picture gives a spritely enough impression of the yacht, which belonged to the Burgomasters of Amsterdam in 1600, running before the wind, her boomless sails goose-winged, a spar bearing out the foresail, her carved and gilded poop soaring high in a curve that the artist may not have grossly exaggerated.

Apart from the official yachts in Holland were the many more owned by private citizens. Holland was rich, with the wealthier enjoying a *douceur de vie* unknown yet in Britain; the elegance of domestic life at this time appears in paintings showing Dutch interiors, such as Van Eyck's Arnolfini, dated about 166 years before Rool's drawing. The highways of the Dutch provinces were the inland seas and intersecting waterways; a yacht was the means of making a door-to-door journey, pleasantly along a waterway instead of over a pot-holed road in a swaying carriage; it was a conveyance adopted by all who could afford it. We see in paintings and drawings some of the grander private yachts of the rich during the seventeenth century; in the mind's eye we have to imagine the great fleet of privately owned craft, ranging upwards from some 18 ft in length, which during the sixteenth and seventeenth centuries were the carriages and daily conveyance of this maritime people.

There are no records of yacht racing or yacht clubs. In all ages except the present, with its plethora of racing dinghies, the majority of yachts have not been racing craft, but cruisers which rarely if ever raced. None the less, the organization of racing and its dominant position as the most spectacular and technically advanced sector in the world of yachting, is an essential feature of the activity as we know it today. Another is the formation of yacht clubs. This would place the origin of yachting not earlier than the latter end of the eighteenth century, and not until the nineteenth was well advanced did yachting in this sense become of social significance.

The first of the few histories of yachting was the well-known, though long out of print, work by the American Captain Arthur Clark, published by Putnam in New York and London in 1904. The work was dedicated to the author's club, the New York Yacht Club, and parts of it may have been written in his beautiful English designed and built cutter *Minerva.* The period considered by Captain Clark in his *History of Yachting* was 1600 to 1815, and he says of it:

'Yachting history may be divided into two eras. The first dates from the year 1600 to the years 1812–15, when The Yacht Club – now the Royal Yacht Squadron – was founded, and modern yachting may be said to have begun; the second from that date to the present time. I propose to deal only with the first, comprising many events of interest, which hitherto have escaped the attention of historians of yachting.'

The 215-year period considered by Captain Clark ends, in fact, at the moment when the activity of yachting was just about to assume a form giving it some relationship to what we know today. He describes yachting as 'the poetry of the sea'. He adds that 'it is amongst the most ancient, as well as the noblest of sports'. But while the lineage of yachting, like some people's family trees, may be stretched back to an improbable past, it is, like most of today's sports and leisure activities, essentially modern in origin.

It is a butterfly of a word to catch and fix. It covers a variety of activities having little in common except that they all take place in the environment of water that is usually – but not inevitably – salt. Yachting may be a country house party staged afloat. It may be the purest athleticism, with a boat taking the place of a racquet, a bat or a sabre. It may be a highly organized and even dangerous operation of team work and endurance, like an ascent of Kanchenjunga. It may be like an anchorite's period of self-discovery on bread and water in a desert. It may be a gentle caravan tour along a coastline instead of country lanes.

Model of a small Dutch yacht of the seventeenth century with the bezan rig, characterized by the very short gaff. Note also the leeboards of the shallow 'shoe' type.

Yachting may be tea on the Squadron lawn or by the Solent or Long Island Sound, with a cocktail party on board a yacht to follow. It may entail a diet of fruit and no smoking in preparation for a series of contests round an Olympic Games course. It may be a transatlantic race with five men on board a yacht no more than 24 ft long. It may be a Friday evening to Sunday night packed on board with the children, pulling ashore for milk in the morning, and dropping anchor in each sandy cove for a bathe. Or it may be days spent catching the flavour of other lands – the sound of Breton voices, the smell of Breton sardines – with evenings following days when everything has gone well. and in the last of the light the yacht swings to an ebb tide that saunters out to sea over darkening mud flats.

In England, it was the splendid evening of Gloriana when Rool's yacht was sailing and yachting was none of these things. Twenty-nine more years had to pass before Henrietta, Queen of Charles I, was to give birth to a baby who – curiously in the light of his later life – was described as very ugly and solemn. The baby had become thirty-one years of age and the new monarch – Charles II – and a little world-weary after the years of exile, when he got up at five o'clock on an August morning in London to see the yacht that had been presented to him by the Dutch, the first English-owned yacht, and purely Dutch in character.

2
Into England

The evening had not yet fallen on the vaulting reign of Queen Elizabeth I – it was the year of the Spanish Armada itself – when a small craft was built for the Queen's personal use. It would be a gross distortion to see in her, any more than in the White Ship, an early English yacht; though the circumstances of her construction would make it appropriate could we do so. This otherwise historically insignificant little craft is curiously suggestive when looked back upon from the present time.

Named *Rat of Wight*, she was built at a small and impoverished fishing village whose cottages like hutches clustered in disorder at the mouth of the Medina river. Its name until lately had been Shamford, and hardly heard of except locally; the new name it had recently acquired fell strangely yet even on local ears. The Queen's father, King Henry VIII, had built here, on either side of the Medina's entrance, two forts, or 'cows', to protect the river which, if it fell into enemy hands, could become a dagger plunged deep into the heart of the island. In time the forts brought Shamford the new name of Cowes, and the one on the western side of the river, enlarged and softened into domesticity, was on a future day to become the home of the Royal Yacht Squadron, the club whose foundation gave Captain Clark the originating date of modern yachting.

'The 28th day we weighed about noon, and anchored thwart Sluis, where came on board us with his yoathes, the Prince of Orange, Count Maurice, with a great train of gallantry and followers, who all lay this night on board the Admiral.'

This was written in 1613 by Phineas Pett, Master Shipwright and Commissioner of the Navy in the reign of James I. His expression here of 'yoathes' is believed to be the first time the Dutch term for yacht appeared in some form of English. But other than appearing in Pett's then unpublished diary, the word continued to remain unknown in England for another half century. The naval architect Sir Anthony Deane told Samuel Pepys, when the latter was Secretary of the Admiralty, that before 1660 'we had not heard of such a name in England'.

This was the year that Charles II returned from exile to occupy the throne of England. When, on 8 May 1660, he was proclaimed King of England in Westminster Hall he was in Breda, and to convey him and his suite to Rotterdam a number of yachts were placed at his disposal. The one in which he took passage appears in a painting by Lieve Verschuier, now hanging in the Rijks Museum, Amsterdam.

The king was impressed enough with the yacht to say that he intended having a vessel like her built as soon as he reached England; and from this remark followed the often recorded events leading to the presentation of the yacht *Mary* to him at the instigation of Van Vlooswyck, Burgomaster of Amsterdam. The yacht was originally built for the Dutch East India Company, and the resolution of the City Council of Amsterdam dated 28 May 1660 records the approval for negotiations to be opened for her purchase. And so yachting reached England.

A year and a half later, on 1 October 1661, we find John Evelyn writing in his diary:

'I sailed this morning with His Majesty in one of his pleasure-boats, vessels not known among us till the Dutch East India Company presented that curious piece to the King; being very excellent sailing vessels. It was a wager between his other new pleasure-boat, built frigate-like, and one of the Duke of York's – the wager 100–1; the race from Greenwich to Gravesend and back. The King lost it going, the wind being contrary, but save stakes in returning. There were divers noble persons and lords on board, his Majesty sometimes steering himself. His barge and kitchen-boat attended. I brake fast this morning with the King and returned in his smaller vessel, he being pleased to take me and only four more, who were noblemen, with him; we dined in his yacht, where we all eat together with his Majesty.'

The yachts mentioned by Evelyn were the *Katherine* and the *Anne*, both English designed and built to compete with the 'Dutchman'. Evelyn mentions that the King's new yacht was built 'frigate-like'. English design was improving, at least for its own conditions, upon that of the Dutch, eliminating their excessive beam and great bluffness and remedying their lack of draught, and with this discarding the Dutch leeboards. The English development in design was towards a narrower, deeper kind of hull with softer underwater lines, and faster, closer-winded vessels were the result. It was in this direction that the Americans were later to move, but far further, in their evolution of the fine-lined schooner.

Two royal yachts racing: Charles II's *Mary*, second of the name, and what is believed to be the Duke of York's (later James II) *Charlotte* immediately astern and up to windward. Both yachts are gaff rigged and without leeboards. Compare this with the Dutch yachts on pp. 8 and 11. Painting by William Van de Velde the Younger, 1680.

The work of designing and building these yachts was largely managed by members of the Pett family, descendants of the above Phineas.

In August 1673 the English-built yacht *Katherine* found herself in Holland after being chased and captured by a Dutch frigate. We owe to this event an excellent description of her by Nicholas Witsen, Burgomaster of Amsterdam, who published his account of the yacht in the second edition of his work on Dutch shipbuilding. It appeared in 1690. The following is a condensation of the translation by C. G. 't Hooft which appeared in the *Mariner's Mirror* in 1919:

'On 4 September 1673, there was to be seen at Amsterdam in front of the palisades a very costly yacht of His Majesty the King of England. It is low in the water, carried ten brass guns, and is so shaped and built that it can keep the sea. Sail, mast and gaff of the shape that is usual in this country (Holland). The stem is upright, after the new fashion, formed of a strong piece. Has a fine yellow painted head supported by sea-nymphs and goddesses; it spreads out wide, but below it is sharp and of good shape, very proper to cut the water and to make good speed. All round on the outside for two feet below the rails it is covered with costly sculpture and gold painted carvings of grotesque figures and plants. It has no leeboards. It is also not tarred outside, as one is accustomed to do to ships here, and this is very elegant, for the Irish timber of which it is made shows all the better its nature and its ruddy colour which is agreeable to the sight; to tar this wood or preserve it with paint or otherwise is not necessary, for it is the finest wood to be found in Europe, and resists all worm and rot. It is true that this wood is not so flexible as the oak that is used for shipbuilding in this country, and therefore this yacht is not of such a curved shape as they are built here,* but more straight sided, which nevertheless does not look bad, though many masters consider this the important feature of a ship. It has low bulwarks. The mast is stepped in such a way that it can be moved to and fro to seek sailing power; from the truck flies a silk streamer which in a calm hangs down to the water; and it is on all sides proudly decked with flags. The lower part of the stern is round, and it spreads out high and broad (after the English fashion).'

Between 1671 and 1677 Charles II had fourteen yachts built at Woolwich, Portsmouth, Rotherhithe, Lambeth and Chatham, ranging in length between 65 ft and 31 ft. They were thus relatively small craft compared with those that led the revival of English yachting in the early Victorian period. But despite the magnet of royalty, the yachting seed brought no impressive yield. The time was evidently not yet ripe for people to take their pleasure on the water. More than 130 years later, in 1812, there are believed to have been no more than fifty yachts afloat, and the Victorian yachting authority Dixon Kemp described these as belonging 'exclusively to noblemen and country gentlemen'. This does indicate the tenuous relationship between the first yachts in Britain and the surge of activity first in Britain and spreading – an amazing crop from the small sowing of 1812 – over the waterfronts of the civilized world. By 1850, a year before the *America* was out in the Solent, there were reckoned to be 500 yachts in British waters, which was still a trifling number.

Before this new and mainly royal sport of yachting disappears from the narrow focus of history, we find in the latter part of the seventeenth century an account of yacht cruising having a remarkable air of modernity.

The Hon. Roger North's biographies of various members of his family were published posthumously in 1842–4. Roger North (1653–1734) wrote of himself: 'Another of my mathematical entertainments was sailing. I was extremely fond of being master of anything that would sail, and consulting Mr John Windham

* It may be questioned that this was purely the reason. The English were designing for their own seaways.

about it, he encouraged me with a present of a yacht, built by himself, which I kept four years in the Thames, and received great delight in her.'

The yacht was smaller than most of those of the reigning king, James II brother of Charles II – open aft and with some accommodation ahead of this; and we may notice yacht-like characteristics of the future in embryo in North's craft, in that she was able to work to windward better than the working craft of the river, while like some of the king's yachts she was ballasted with lead, a luxury that did not become usual in yachts until the latter half of the nineteenth century. Roger North sailed down-river as far as Sheerness and Harwich, and of one day afloat he said: 'I, with my frind Mr Chute, sat before the mast in the hatchway, with telescopes and books, the magazine of provisions, and a boy to make a fire and help broil, make tea, chocolate, etc. And thus, passing alternately from one entertainment to another, we sat out eight whole hours and scarce knew what time was past. For the day proved cool, the wind brisk, air clear, and no inconvenience to molest us, nor wants to trouble our thoughts, neither business to importune, nor formalities to tease us; so that we came nearer to perfection of life than I was ever sensible of otherwise.'

Here indeed is the voice of the dedicated cruising yachtsman. Another voice too is familiar. It will be noticed that Roger North writes of sailing as being one of his 'mathematical entertainments'. The phraseology may have an archaic note; the sense is quite modern. Throughout the period of modern yachting there has always been a proportion of technically and scientifically inclined yachtsmen whose amateur work has been influential in the design, construction and handling of small craft. Such men have been able to bring to the creative and scientific aspects of boat design better educated, more open minds than had ever before been applied to matters of technique in such small craft.

Roger North died in 1734, and his yacht cruising, like the former king's yacht racing, does not appear to have charmed enough people to leave much legacy from these first examples of sailing for pleasure in English waters. The general attitude was epitomized by some friends of North in Harwich, who asked him and his companions when they arrived in the yacht whether 'we had left our souls in London, because we took so little care of our bodies'.

While the yacht cruising of Roger North strikes a modern note, and equally the racing of the king and his brother, this first phase of English sailing for pleasure – its yachts built to resemble warships and almost all of them entered in the Navy List – remains a dim image of the yachting that was to emerge in the nineteenth century. And had it not been for the diaries of Samuel Pepys and John Evelyn little today would be known of the seventeenth century's royal yachting. We may wonder how much has fallen by the wayside along the road of history about other craft that may have been used for amateur sailing during the eighteenth century. Not much, we may suspect, though no doubt a little that would help to bridge the gap between the yachting of the Stuarts and of the later Georges. The novelist Henry Fielding, author of the novel *Tom Jones*, provides a scrap of information, though of a negative kind.

Anyone during the eighteenth century going down the Thames would have seen on either hand many stretches of mainly green shore dotted with windmills, farms and houses set in parks, with the delectable river hardly used for pleasure, even by those living hard by it. Fielding made a voyage to Portugal in 1754 in the hope of recovering his health. In Lisbon he died. Later there was published his *Journal of a Voyage to Lisbon* in which, writing of the Thames below Greenwich, and the fine houses on either bank, he said:

'And here I cannot but pass another observation on the deplorable want of

Opposite

Charles II's yacht *Royal Escape*, classed in a list of the Royal Navy as a smack of 34 tons. Length on the keel 30 ft 6 in. Painting by William Van de Velde the Younger.

taste in our enjoyments which we show by almost totally neglecting the pursuit of what seems to me the highest degree of amusement. This is the sailing ourselves in little vessels of our own, contrived only for our ease and accommodation to which situations our villas, as I have recommended, would be convenient and even necessary. This amusement, I confess, if enjoyed in any perfection, would be of the expensive kind; but such expense would not exceed the reach of a moderate fortune, and would fall very short of the prices which are daily paid for pleasures of a far inferior rate. The truth, I believe, is that sailing in the manner I have mentioned is a pleasure rather unknown or unthought of than rejected. . . .'

But such ideas, trite enough to the mind of the twentieth century, were outlandish to those of the eighteenth. Dr Johnson saying that 'he who goes to sea for pleasure would go to Hell for a pastime' spoke for the century. The last two Stuart kings may have sailed for the joy of it, but plenty of people in eighteenth-century England no doubt felt that they were just the sort of people who *would* have gone to hell for a pastime. For the majority the sea, even London's winding, beguiling river, was made for work not play. It needed a world of macadam and chimneys, steam power and steel, to rise where the fields, country towns and teams of horses had once been, before the average person contracted the urban *malaise* and could feel any emotional complicity with the Water Rat who claimed 'Believe me, my young friend, there is *nothing* – absolutely nothing – half so much worth doing as simply messing about in boats. Simply messing,' he went on dreamily, 'messing – about – in – boats; messing. . . .'

It was not until 1908 that Kenneth Grahame put these words into the mouth of Water Rat.

But there is other evidence of about the same period as Fielding's notes to indicate that yachts had become a recognized if not common type of small craft. Fredrik Henrik Chapman, sometimes if inaccurately, known as the 'father of naval architecture', published his *Architectura Navalis Mercaturia* in Stockholm in 1748. The series of beautifully drawn plates in this work has maintained its value for 200 years. A part of Chapman's purpose was the systematic classification of ship types, and he was collecting information all over Europe during some twenty years prior to the publication of *Architectura Navalis*. His enthusiasm led him to a brief spell in an English prison for being in possession of what nowadays might be described as 'classified material'.

The ships and craft so magnificently illustrated in the plans are thus representative of those sailing during the latter half of the eighteenth century. Eleven yachts are illustrated in the section of plates entitled 'Vessels for swift sailing and rowing'. One, described as a 'sloop or yacht' is 40 ft in length between perpendiculars and displaces 25 tons. Another, a 'frigate or pleasure vessel for sailing', is 77 ft in length. Two 'schooners, or pleasure vessels for sailing' have proportionately narrower and less deep hulls than the others, and were equipped with rowlocks for six pairs of oars.

One of the smallest yachts illustrated is particularly attractive. Only 26 ft 1 in in length between perpendiculars, the design gracefully prefigures the many small cruising yachts which, more than a century hence, were to become so common; though none more elegant than this predecessor of them all. She is beamier and has less draught than most of the later cruisers of her length, in which she suggests the American rather than the British tradition in yacht design, but it is the Dutch influence that guides her shape; indeed, the design may have been based on a Dutch yacht of the period. There is a cuddy right aft, a cockpit extending to just short of amidships, and then a coach roof (or cabin top) extending over the

An English 'official' yacht; Dover Harbour *c.* 1800 with the customs yacht on the right.

amidships portion and covering a cabin 6 ft long but rather deficient in headroom – only some 4 ft 6 in under the beams of the coachroof, giving sitting headroom. Ahead of this is a long fo'c'sle with even less headroom. This general scheme of layout is found in most of Chapman's yachts. In one 43 ft 5 in length between perpendiculars the cabin amidships is 11 ft in length and has standing headroom.

All these craft have generous beam compared with later European practice. They have bluff round bows, and the larger of them have the main accommodation concentrated aft in the poop, a practice that was to be revived in some yachts two centuries later. Chapman's small collection of yacht plans throws a brief shaft of careful light on the yachts of the eighteenth century which otherwise have left such scanty records of their character and activities.

Such craft as these were in existence when the Water Club's fleet was out in Cork harbour, and later when the Cumberland Fleet appeared on the Thames.

Fielding was presumably unaware, when he made his last journey down the Thames, that even then there had already been in existence for thirty-four years a yacht club in Ireland, the first in the world of this kind of institution.

In June 1970 a fleet of yachts sailed from the Brenton Reef Light Tower on a race across the Atlantic to Cork Harbour. It was sponsored by the Cruising Club of America and the Royal Cork Yacht Club to commemorate the 250th anniversary of the latter.

Queen Anne was dead and a rather raw German king was on the English

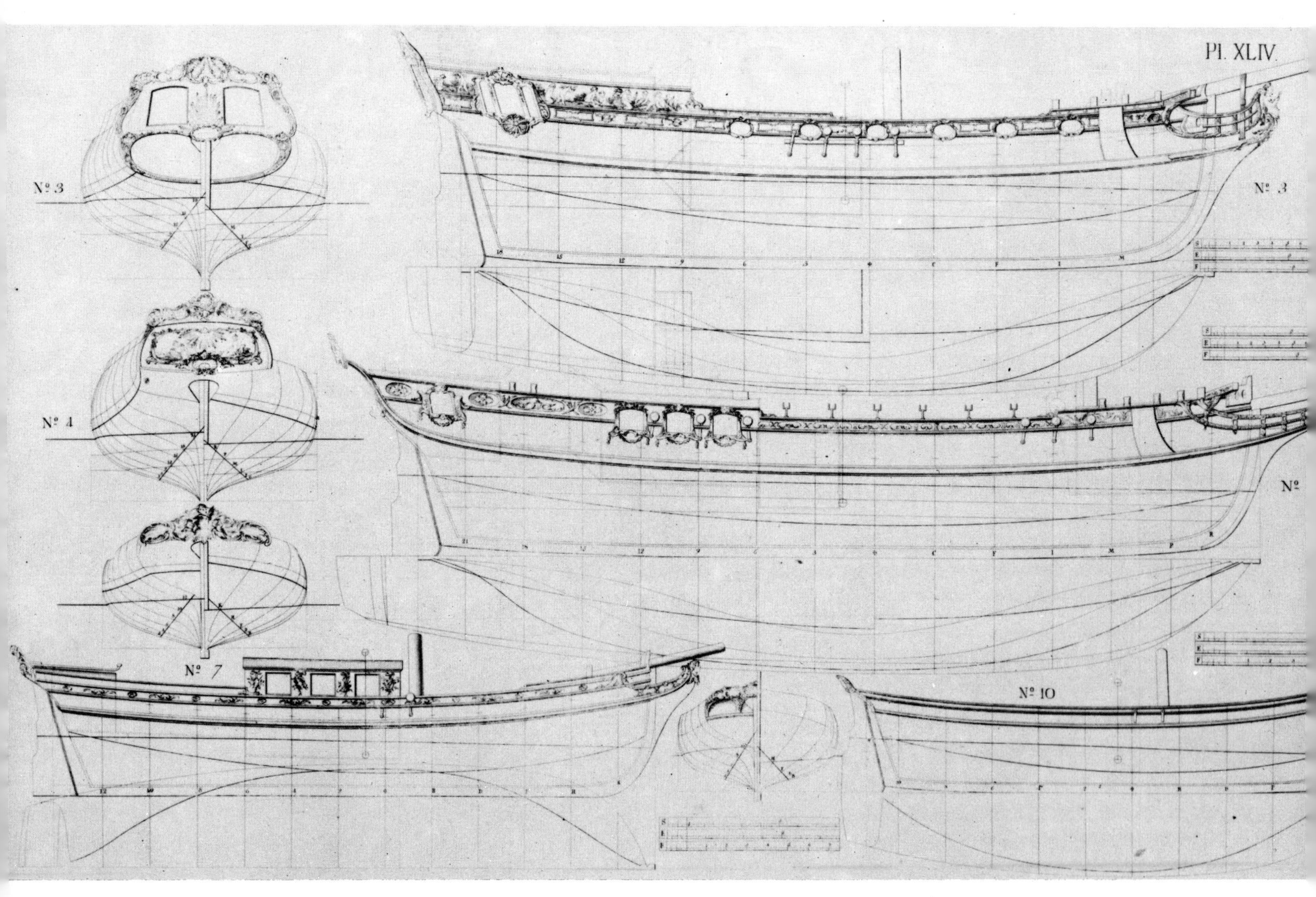

Four yachts shown on a plate in Fredrik Chapman's *Architectura Navalis Mercatoria* (Swedish edition) first published in 1768. They range in length between 54 ft (length between perpendiculars) and 24 ft. The latter, at bottom right, is a sturdy little boat with 7 ft 10 in beam and 2 ft 6 in draught.

The Water Club of Cork putting to sea in 1720. The ritual of fleet manœuvring and exercising in company, not racing, is portrayed here; the stately maritime minuet. There are no sprit-rigged yachts in the fleet; thus early the rig disappeared from British yachting.

throne when on Haulbowline Island, set in the magnificent expanse of Cork's land-locked harbour, a flag was raised – that of the newly founded Water Club of Cork. It was simultaneously a yacht club, though not yet called such, a coast guard and a cruising association. A new kind of organization came into being, not in Holland where yachts had received their name, or in England where a new role for such craft had been anticipated, but in one of the finest natural harbours in the world situated in what was once known as John Bull's other island.

The flag raised amid such fine coastal scenery in 1720 was perhaps a Union Jack with an Irish harp and a crown in the centre. Meetings of the club were held once every springtide from the first spring in April to the last in September; and those Irish gentry of the early eighteenth century limited their numbers to a mere twenty-five; while stringent rules governed the introduction of guests, and sartorial matters were not neglected: '. . . no long tail wigs, large sleeves or ruffles to be worn by any member of the club.'

Six years before Fielding had deplored the yachtless Thames two English tourists had described the proceedings of the Water Club of Cork. They were 'something like that of the Doge of Venice's wedding the sea. . . . Their Admiral, who is elected annually, and hoists his flag on his little vessel, leads the van and receives the honour of the flag. The rest of the fleet fall in their proper stations, and keep their line in the same manner as the King's ships. The fleet is attended by a prodigious number of boats which, with their colours flying, drums beating, and trumpets sounding, forms one of the most agreeable and splended sights. . . .'

Thus it was off Haulbowline Island, by the deep-green shores of Cork Harbour at the times of the summer springs for some forty-five years, during which time no other yacht club existed. Then the gloriously gilded little fleet, the firing guns, the trumpets and the beating drums vanish for a time from history, along with the captains and admirals, the Knights of the Island and all the massed

The Cumberland Fleet racing by St Paul's in 1782. In this up-river scene the craft racing are small. The cutter at anchor in the foreground illustrates the general type.

flutterings of bunting, leaving Haulbowline quiet as it had been before the summer times of 'live water'.

What the sudden storm was that dissipated the Water Club's gay fleet is unknown. Between 1765 and 1806 there are no records. When in the latter year it was re-formed there was another yacht club in existence, beside a Thames less green than the one Charles II or Fielding had seen, though still green enough. This was the Cumberland Fleet. Meanwhile the Water Club in Cork harbour had an uneven existence, becoming humbly known at one time as the Little Monkstone Club, until it was re-established on a firmer basis in 1828 under the title of Cork Yacht Club. It became a royal club.

The brief account of the Water Club's early days reveals characteristics of yacht club activity which were to disappear as yachting developed. The sailing and manœuvring in company, the evolutions performed by signals, the gunfire and salutes and careful etiquette, all the drilled steps in a nautical minuet, were the apings of naval procedures expressing the yachtsmen's feeling of affinity with the navies of sail. This affinity was to remain, though taking less picturesque if more useful forms. For a while yachts were regarded as a means of improving the breed of the smaller types of naval vessel. Later, when navies were composed of steam-driven steel ships, only the link of the sea remained between the yachting and fighting fleets. But the link remained, and the amateur and professional seamen of the yachting fleet became the navies' valuable reserve of officers and men in the warfare-dominated semi-civilization lying ahead.

William Falconer, in his *Universal Dictionary of the Marine* first published in 1769, defined a yacht at some length, chiefly emphasizing its royal and official uses: 'A vessel of State, usually employed to convey princes, ambassadors, or other great personages from one kingdom to another.' But he did mention briefly that they might be 'used as pleasure boats by private gentlemen'.

Model of Commodore Taylor's second *Cumberland* of 1790.

Such, for example, as one owned by Lord Ferrers a few years after the appearance of the Dictionary. In one respect this craft typified the best yachts of the future in her remarkably fine sailing performance to windward. Her performance was described with something close to wonder in the *Gentleman's Magazine* in 1773. Working to windward in the Thames she drastically outsailed at least a hundred working craft of sail, and part of the time against a strong spring ebb performed with an ability, the writer said, such as 'was never done before, nor can be done by any other vessel.'

As the present author wrote in his *A History of Seamanship*: 'The battle for weatherliness, which we have seen to have been a constant preoccupation during the ages of sail . . . is one in which physical laws prohibit a victor. But at least the modern yacht is able to put up a much superior fight against a head wind and sea than even the most respected working craft of the past, such as the smart luggers of the Brittany coast, the pilot cutters of the Bristol Channel, North Sea, or of the Chesapeake, the best lateeners of the Mediterranean, or even the fishing schooners of the Grand Banks, some of which were built for racing and were essentially yachts.' In 1773, when Lord Ferrers was impressing the sail-crowded Thames with his yacht, which conformed with what was described as 'his new method of constructing ships', the yacht as a type was at the beginning of its career of progressive refinements which were to lead to the most efficient of all wind propelled vessels ever devised by man.

The pressure behind this growth of efficiency was the sport of racing, which was first practised in a form recognizably modern in the 1870s. It was at this time that the term 'regatta' travelled from Venice to England; and the Duke of Cumberland, a brother of King George III, did something towards establishing a reputation for the nautical, which was not strong in the Hanoverian monarchy, by encouraging yacht racing. In 1775 there was formed the Cumberland Sailing Society, an organization that continues today in the Royal Thames Yacht Club, one of the most powerful yacht clubs in the Britain of the 1970s as it had been one of the first 200 years earlier. In 1775 the Water Club of Cork was passing through its lacuna phase before re-emerging as the Cork Yacht Club in 1828. There was, however, in existence another yacht club in the British Isles when the Cumberland Sailing Society ran its first races, the Starcross Yacht Club in Devon, two years old at the time, which has passed more quietly and with less distinction through history and survives today.

The boats involved in that early Thames racing were small and cutter rigged. Thomas Taylor, who became Commodore of the Cumberland Fleet in 1779, owned a number of yachts famous along the waterfront Thames in their day and now recalled in many prints hanging on club walls. His *King's Fisher* was a sturdy clinker planked craft no more than 20 ft in length and 7 ft beam. He subsequently owned four yachts named *Cumberland*, of larger though still modest size. But those Thames craft had deep draught and great bulk for their length; they carried running bowsprits more than half as long as the hull and spread broad topsails above their long gaffs. The fourth of Taylor's *Cumberland*s is noteworthy in the light of future yachting history in that she was fitted with three centreboards, or sliding keels, a device introduced from the U.S.A. Not only were the boats not large but the Corinthian spirit was strong in their handling. Owners had to sail their own craft when racing, rules sometimes specifying that they might not have the help of more than two crew.

But life in the Cumberland Fleet was not all hard sailing. There were also the bumpers being passed round in gay Vauxhall Gardens, the huzzas of the crowd on the banks, the colours flying, the music playing. Nor was the racing always conventional by later standards. Two yachts became foul of one another during a Thames race in 1795, whereupon the skipper of one set about the rigging of the other with a cutlass, causing a good deal of damage. The contretemps appears to have been between the two leading yachts; after this it was the third boat that won. While the racing occupies most of the annals of those Thames sailing days, there was far more cruising. There may not have been more than two or three races in each season, while the yachts sailed regularly during the summer along the reaches of the river between Vauxhall and Sheerness, whose charm was so soon to be wreathed in the twentieth century's steam and smoke, all graciousness sacrificed to create wealth, which amongst much else supported the future of yachting.

The Duke of Cumberland was patron of the club from its foundation until 1790, but he does not appear to have taken part in its activities after 1782. On the coronation day of King William IV in 1823 the club's name was changed to His Majesty's Coronation Sailing Society, but this did not last long. A matter of days later a race produced a robust altercation which resulted in a club rupture; the Thames Yacht Club was founded as a separate body which was in 1831 to absorb the Coronation Society, thereby assuring that the Thames Yacht Club became the true descendant of the Cumberland Fleet. A year prior to this the Thames had become the 'Royal Thames'.

There was thus regular yacht racing on the Thames during the period 1775 and 1815, at which latter period the Yacht Club, later the Royal Yacht Squadron, was founded, which in due time was to establish organized yacht racing in the Solent. But it was the Thames, not the Solent, which fostered the sport in its modern form. The yachts ranged in size between 4 and 17 tons – tiny compared with the great craft so soon to appear on the Solent, where the Squadron was later to set a lower limit of size for the yachts of members at 30 tons. Time allowance appears never to have been used on the Thames, as indeed for some years to come it was neglected on the Solent also. The report of a race in July 1807 for the much sought Vauxhall Gardens Silver Cup shows the result of this: 'There was a stiff breeze from the southward, which occasioned the *Bellissima* being so much heavier than the others, to carry a great press of sail and enabled her to keep the lead the whole distance without the least chance of the others coming up.'

During this period there are to be found brief contemporary reports of racing at other places round the coast, specks of light on a scene historically dark, indicating the existence of organized local sailing, perhaps mainly between working boats manned by those who were soon to find themselves during the summer months of the coming years the professional hands of the yachting fleet yet to be. We hear of racing at Bristol and Southampton, and of a Dutch yacht being involved in the sport in the latter waters. In 1805 there was a revival of the Water Club of Cork, and as in 1720 sartorial matters were again of concern; though now they had a more modern note than the stipulations about long tail wigs and ruffles of the earlier date: now it was ruled 'That the wives and daughters of members of the club, be also considered members of the club, and entitled to wear their uniforms.'

The seaside had become fashionable and there was growing up a medical belief that even total immersion in seawater might prove beneficial. Desirable residences were appearing in Cowes, and amongst the working boats that occasionally raced and were always busy, were to be seen a small number of gentlemen's pleasure vessels. It was also a period when the old tradition of clubs was being revived to produce the flourishing crop of Victorian clubmen. In 1815

Cowes Castle as it appeared in 1801. The Royal Yacht Squadron did not move in until 1858. To the left a brig sails eastward down the Solent towards Portsmouth, with the Hampshire shore behind.

some forty-two gentlemen with a liking for Cowes founded at a meeting in London the Yacht Club. It is likely that this was the first time that the word 'yacht' appeared in the title of such an organization. Indeed, the word may have been little used among the members of the Water Club or Cumberland Fleet; its appearance would suggest that the aura of Solent yachting rose early, that 'Sacred Cowes' was already burnishing its censers. It was stipulated that to become a member of the Yacht Club ownership of a vessel of at least ten tons was required. Amongst the original members were two Marquises, three Earls, four Viscounts, four Barons and five Baronets, while those unblessed with a title at least bore well-quartered arms. It is not surprising that in the world of the early nineteenth century a club with such a membership should be swept ahead on a wave of coronets. However, in the first volume of the club's history it is written with becoming modesty that 'the greatness to which the Yacht Club did not feel itself born was suddenly thrust upon it'. In 1817 the Prince Regent, heir to the British throne and soon to be King George IV, gave notification that he wished to become a member. He was of course appreciatively accepted, and two of his royal brothers were soon also applying to join. When the Regent became King in 1820 he consented to the club becoming known as the Royal Yacht Club – the first of all 'royal' yacht clubs. It was now forty-five years since the King's brother the Duke of Cumberland had given his name and patronage to the Cumberland Fleet. In 1820, as noted above, the name was changed to the His Majesty's Coronation Sailing Society; but there could be no question that the much younger club at Cowes had overtaken it in prestige, though the latter did not yet organize any racing. Five years later the Royal Yacht Club acquired a premises in Cowes on the Parade. It did not move into the Castle it now occupies until 1858, seven years after the *America* had come to England. Meanwhile, having become Royal it raised its yacht tonnage qualifications to twenty. This was later to become thirty.

It has been estimated that as late as 1840 there were not more than some one hundred yachts exceeding 20 tons afloat in British waters. Their design followed that of the smaller naval craft – the brigs, schooners and cutters – and the cutters of the Revenue Service. In the heaviness of their sparring and the crush of canvas they set, the latter represented the furthest limit of sail carrying that had yet been reached with the fore and aft rig. The Revenue cutters have been described as the father of the modern yacht, though in fact the type was to be transformed in the next few decades. The yachts, with midship sections as round as an apple; the run of their hulls lines fore and aft bluff forward and relatively fine aft, known as 'cod's head and mackerel's tail'; their ballasting, of iron pigs carried internally, sometimes mixed with stone – these characteristics were of the older tradition in design which was to be transformed by the yachting of the next decades, while the smaller commercial and naval craft under sail were ousted by steam.

Initially yachts were expected to provide an example for commercial, and particularly naval craft. In their privately printed history of the firm, Samuel White of Cowes, the story is told of an early famous yacht:

'When the *Waterwitch* was built the naval twelve-gun brigs were bitterly criticized; their popular name of "coffin brigs" is sufficient indication of the reason. To get the desired speed with their length and lines it was necessary to over-rig them terribly and any number of them came to grief, while the French wars had proved that even that dangerous feature did not give them the speed that was desirable. Lord Belfast commissioned White to design and build a brig

When yacht clubs tried to fulfil their expressed object of encouraging improvements in naval architecture . . . Lord Belfast's Cowes-built brig *Waterwitch* on trials with brigs of the Royal Navy, 1834.

which could carry a battery just as heavy, or heavier, than the standard, which should be faster and more weatherly and, above all, safer. Noble yachtsmen were then welcome to join company with naval squadron in evolutions, but the *Waterwitch* soon outwore her welcome for she beat the best of the official designs time and again. Eventually, after stubborn opposition, the Admiralty was forced to buy her . . . Lord Yarborough's brigantine yacht *Kestrel*, built in 1845, was also bought by the Royal Navy.'

The yearly period of yachting in both the U.K. and U.S.A. is bound up in the maritime traditions of the two nations. For a time it was a decorative and also potentially useful extension of the larger maritime affair. In 1850 Donald Mackay was only forty years old. The *Great Republic* was not launched until 1853. In 1859 the *Victoria* was launched for the Royal Navy, the last wooden First Rate three decker, direct descendant of Nelson's *Victory*; she was still flagship of the British Mediterranean fleet in 1867 and the handling of sails was still the most sublime aspect of the naval career. The famous clipper race, ending in what is usually described as a dead heat, between the *Ariel*, *Taeping* and *Serica*, occurred in 1866, and the Suez canal was not opened until 1869. Hence, in Britain, the early yacht clubs would state amongst their objects the improvement of naval architecture, and there followed official recognition and royal patronage. Sometimes the clubs disappointed in this respect. The Royal Yacht Squadron early in its life was taken to task in the publication *Sporting Annals*:

'The naval exploits of this club appear to be represented in a way little calculated to perpetuate their celebrity or to further some of the principal objects which ought to be connected with the institution. We have the right to presume that where so much wealth and splendour are wasted in the production of first-rate specimens of naval architecture, the great object of so much competition must be to excel in the art of swift sailing, and yet we can gather no information upon the subject from the late evolution of the grand fleet (e.g. yachting fleet) at Spithead. . . .'

As soon as yachts began racing in competition more or less serious, the problem arose of how to classify them; for while all may be fair in love and war, it is widely conceded that in sport like should meet like and win by superior skill only. You do not set a lightweight to fight a heavyweight. But what standard should be used for classifying yachts? How big is a yacht? How fast? Is she as big as she is long, as big as she is broad, or as big as she is heavy? How long is a piece of string?

During the years of seafaring that preceded yacht racing efforts to establish the size of ships had been made in various systems of tonnage measurement which determined the harbour dues that should be paid. The principle was that the bigger ship paid more than the smaller. It is also a fact, though it was not fully grasped for a long time, that all else being equal, the bigger ship is innately faster than the smaller. It was therefore not unsound in principle, though much oversimplified, when in the early years of yacht racing the tonnage laws used in the commercial world should have been adopted for classifying yachts.

The tonnage laws, however, were devised to assess bulk, the amount that a ship could carry, rather than the speed at which she was capable of sailing. It was soon found that the stress of racing competition made commercial tonnage laws unsuitable for the assessment of potential speed. The age opened of rules devised specially for classifying racing yachts, and these rules have ever since been the thread running continuously through the story of yacht racing. They have determined, at any moment, the types of yachts that are successful. They, not the great natural forces of wind and sea, have been the prime influence on the character of racing yachts. Down the years these measurement or rating rules, have become more complex until, by the 1970s, a rating rule might need to contain in it a stout volume of close print, a dictionary of definitions, and a battery of formulae capable of being mastered only with the aid of computers. It was all still relatively innocent, if not all light, even in 1892, when Lord Dunraven, who had not yet challenged for America's Cup said of rating rules:

'With shame I confess that the problems and calculations, the combinations of straight and crooked lines, with large and small numerals and Latin and Greek letters, the mathematical contortions and algebraic hieroglyphics . . . are meaningless to my uncultured eyes. They are fascinating; I admire their beauty, and can well understand that inventing rules for rating must be the most charming pursuit for intellectual yachtsmen. . . .'

In relation to matters as they were at the time Lord Dunraven was writing, as he often behaved, for effect rather than with detachment. Rating rules were still very simple indeed compared with what they were to become, though his attitude was that of the average yachtsmen, for whom the problems of yacht measurement were too esoteric to appeal. But as the British yacht architect George Watson said, at about the same time as Dunraven was writing to *The Field* 'Throughout the modern story of yachting the tonnage question has been the all-absorbing one.' Historically, it has to be faced that much in the sunny, salty story of yacht racing will remain obscure, and the qualities and faults of the yachts that acted in it be hidden, unless the ever pervasive, ever changing influence of the governing measurement rules be taken into account.

It will shortly be noted that when the *America* won her race in 1851 no allowance was made for the size of the yachts involved. At that time the rule of measurement used for British yachts was almost contemporary with the first yachts that had been sailed outside Holland. It was derived from the first Act of Parliament that had set ship tonnage on a rational basis, in 1694 as amended in 1773, a couple of years before the Cumberland Fleet had come into existence. It was due to the Cumberland Fleet, in its later guise of the Royal Thames, that the amended rule of 1773 was modified and became, in 1855, the one known as Thames measurement. It may seem odd that as late as 1855 yacht measurement in Britain was derived from a rule formulated in 1694; it is perhaps even more odd that the Thames tonnage of 1855 is still used in the 1970s and recorded in Lloyd's Yacht Register, though it was long since discarded for the rating of racing yachts.

The effect of the Thames measurement rule was to encourage very narrow yachts, for by reducing beam it was possible to increase length whilst retaining the same tonnage. The effects of this will be seen below.

Tonnage rules, such as those described above, might measure the size of yachts. To compensate for the disabilities of the smaller craft they must then receive a handicap or an allowance of time in some proportion to the tonnage. The racing in the Solent during the decade of the 1820s effectively demonstrated the advantage of large yachts, and it was a few big expensively maintained cutters that dominated the racing. Three superb yachts became famous, the *Louisa*, *Menai* and *Lulworth* belonging to deep-pocketed owners who poured out money on their craft. It became evident in the strenuous racing that *Louisa* and *Menai* of respectively 162 and 175 tons were generally faster, if there were any weight in the wind, than the *Lulworth* of 127 tons. Yachts smaller than this trio had little chance of the cups. In 1830, *Lulworth*'s owner, Joseph Weld, had the *Alarm* built, a cutter of 193 tons, which was able to turn the tables on *Louisa* and *Menai* and in her turn dominate Solent racing by largeness.

It was during the 1820s that a primitive system of time allowance was introduced on the Solent. In 1829 the fleet racing for the Cowes Town Cup was split into six classes. The yachts within each class gave no time to one another, but each class received time from those larger. Subsequently the fleets were divided into the same number of classes, and yachts raced only against the others in their own class. This established the important principle that racing should be between yachts within a small range of sizes.

Joseph Weld's large and famous cutter *Lulworth* (left) in 1828 entering Portsmouth harbour with her square yard lowered. A contemporary wrote wonderingly of her that he believed the mainsail contained 1000 yards of canvas. An unknown schooner yacht is also shown beating out of the harbour.

The time allowance question, like that of measurement, is also a continuous thread running through the story of yacht racing. It has to be appreciated that no system of time allowance can be perfectly just. It is a complex technical matter which may be summarized briefly: methods of allocating time allowance rest upon the assumption, which is not true, that a boat of one size will always bear a certain and only slightly variable relationship in speed to a boat of another size. In practice in very light winds a yacht half the size of another may be the faster, but as the wind and speed of both increases the proportion of the small boat's speed to the bigger decreases. Since the fallacy inherent in time allowance has a less potent effect if the allowances are not too large, a guiding principle of handicap racing eventually emerged, that in any class the range of yacht sizes should not be too great. The frequent failure to observe the principle produced periods of poor racing. There is record of a race organized by the Royal Yacht Squadron in 1841 in which the eleven entries ranged between 31 and 393 tons. No system of time allowance could rationally make adjustment for such disparate sizes and speed potential.

While imperfect or no handicapping systems led to a hopeless position for the smaller yachts and a legion of disgruntled owners, objections were occasionally raised to the basic philosophy of time allowance. It was considered not only unjust but derogatory to the evolution of the best yachts that the finer and larger amongst them, which crossed the finishing line first, should have to sacrifice their prizes to smaller and perhaps inferior vessels, 'wretched craft' one incensed writer called them 'not worth half the value of the Cup. . . .' Sailing Committees were facing an intractable problem with a scientific element in it which hard-bitten sailing men were rarely competent to handle. But stumblingly, the modern method of yacht racing was being evolved. In 1841 yachts racing under the Royal Yacht Squadron rules were started together and the time allowance was applied at the end of the race to their finishing times. Two years later there was introduced what became well known in later years as the Ackers' Graduated Scale, devised by G. H. Ackers, a member of the Squadron.

The first recorded race of the New York Yacht Club, sailed in 1844, was conducted with a time allowance the effect of which was to give each yacht about one-eighth of a second per ton (Custom House measurement) per mile of the course. The allowance was applied at the beginning of the race, yachts starting from anchor, the smallest first, followed in order of size by the others according to their handicap. Tonnages ranged, in the fleet of nine, from 17 to 45 tons, and the three yachts which completed the course were separated by only forty-five seconds. But the first two of these were the two biggest entries; and the race was unsatisfactory in many respects. Misunderstood sailing instructions and yachts sailing the wrong course are one of the commonest misadventures in yacht racing and were present in this one, as they have continued to be ever since.

The patriarchal trio of the Royal Cork and Royal Thames Yacht Clubs and the Royal Yacht Squadron was followed during the later 1820s and 1830s by a lively succession of club creations which placed centres of yachting round the coasts. Some of these, such as the Royal Eastern, founded in Edinburgh in 1835, failed to survive. A few of those that remain in existence, are still active leaders in the world of yachting today; others, a little somnolent now, a little like the back-wood peers of an ancient aristocracy, are today's links with the exciting formative period of yachting. Of those still remaining, the Royal Dee was founded in 1815, in the same year as the Royal Yacht Squadron. There followed the Royal Northern (1824); Royal Gibraltar (1829); Royal Irish (1831); Royal Southern (1837); Royal London and Royal St George (1838); and in this year there was also established the Royal Hobart Regatta Association. The styles of these clubs are given in their modern form.

With the appearance of the Royal Dee there arrived a club on the west coast to match the one – and yet one only – on the south. With the Royal Northern the River Clyde began to emerge as the beautiful centre of Scottish yachting. Originally there were two divisions of the club, one in Belfast, the other with its clubhouse in Rothesay. The former was dissolved in 1838. The Clyde division prospered as this river became in later years and until eclipsed by the Solent the principle home of yacht building in the grand manner. To be Clyde, rather than Solent, built was for a period the proudest claim for yachts, steam or sail. King William IV became Patron of the club in 1830, when it received the style of 'Royal'.

The Gibraltar Yacht Club was founded in 1829, and thus we see, in the earliest period of yachting, its first reachings-out over the world. No yacht club survives in the U.S.A. from this date, or indeed from more than a decade later, and it is doubtful whether any existed. In 1837, we have seen the Royal Southern Yacht Club was formed, its first clubhouse, an elegant building still standing by the Royal Pier by Southampton Water. Then Southampton Water had the graciousness that the old walled town itself had also not quite lost in the rush of port and railway developments, and it was possible for a handsome fleet to moor just offshore. It is a happy demonstration of the continuity of things that the club, moving in 1938 to the Hamble river, which runs into the Solent near the entrance to Southampton Water, finds itself still at the centre of southern yachting, its clubhouse almost trailing the front doorstep into that river dense with yachts between its broad estuary and the bridge at Bursledon.

The Royal London was formed a year later. Known subsequently as a Cowes club, its premises now ideally situated on the Parade, it began its days organizing races on the Thames above bridge. Its activities extended down-river and out to the Nore, a move considered more daring then than it would today. In 1882 a branch clubhouse was opened at Cowes, a London clubhouse being retained. The growth of this club holds a mirror to that of all yachting during the latter half of the nineteenth century. In 1848 there were fifteen boats in the club fleet with an aggregate tonnage of 161 tons and an average size of 11 tons. In 1892 there were 248 yachts with an aggregate tonnage of 17,000 tons and an average size of 69 tons. It was in the same year that the Royal St George was founded, and yachting took root at another point on the Irish coast, in Dublin Bay.

It is appropriate to note here that twenty-two years after the establishment of the Royal Northern there was formed in Glasgow what originally called itself the Clyde Model Yacht Club. The term 'model' in its title is indicative of the then prevailing attitude to what a yacht should be. The new club had as its object to further 'a greater amount of emulation among the proprietors of small yachts' – a significant step taken as early as 1856, five years after the *America* had first sailed in English waters, towards the future character of yachting in small craft that was to become dominant a century later.

Since the smallest class of yacht acknowledged by the Royal Northern was 8 tons, the Model Yacht Club accepted yachts only of under this size. There followed racing for what, by the standards of the time, had become regarded as tiny craft, there being as well as the 8 tonners classes for 6 tons and under, and 4 tons and under; also for boats of 19 ft overall. It appears from the records that this bold step towards the future met with little response. In later years the club broke away from its original tonnage bounds and becoming simply the Clyde Yacht Club it settled at the heart of the river's yachting, at Hunters Quay. By1872 there were yachts of over 100 tons in its fleet, though also ones of 5 tons; in 1872 it became a Royal Club.

Detail from William Burgis' print of New York Harbour, 1717, with, on the right, a sloop – or what in Europe would perhaps have been called a cutter – described as 'Colonel Morris' yacht *Fancy* turning to windward.'

3
To the New World

Captain Arthur Clark wrote of Henry Hudson's *Half Moon* that she 'was in all probability the first European vessel – and certainly the first yacht – that ever passed the land now known as Sandy Hook'.

It was 3 September 1609. The *Half Moon* had left Amsterdam early in April and anchored somewhere on the coast of Maine on 12 May. She was, we have seen, in the terms of the day a 'yacht', belonging to the Dutch East India Company; from which the conclusion might be drawn that there was yachting in the U.S.A. before it appeared in Britain. For Charles II and his yachts did not reach England until 1660. We are faced again with the question of what is a yacht? It would certainly be stretching the point to regard Hudson and his valiant crew of sixteen as yachtsmen.

But it is not stretching the point to claim that in the New World, as in the old, the first yachtsmen were Dutch. During the latter half of the seventeenth century the Dutch settlers of New Amsterdam would fit out the narrow quarters of small craft built for trading with touches of comfort, and like their families by the Sheldt and the canals of home sail occasionally for pleasure on the Hudson and Long Island Sound. Also, as for those at home, the small craft was an important means of transport, and an obvious convenience of life was to own such a means privately. But while there are historical records of this in seventeenth-century Holland there are none for the early days of New Amsterdam.

But the famous print published by William Burgis of New York Harbour in 1717 shows several yachts, and a most revealing observation on this period appears in the *Memorial History of New York*: 'Racing on the water was not much in fashion, though the gentry had their barges, and some their yachts or pleasure sail-boats. The most elaborate barge, with awning and damask curtains, of which there is mention, was that of the Governor Montgomerie, and the most noted yacht was the *Fancy* belonging to Colonel Lewis Morris, whose Morrisania Manor, on the peaceful waters of the Sound, gave fine harbour and safe opportunity for sailing.'

This yacht *Fancy* appears in the Burgis print, and is described as 'turning to windward with a sloop of common mould'. She is thus clearly differentiated from a working boat, though she is not dissimilar in general type; and it will also be noted that in the *Memorial History* yachts and pleasure sail-boats are put into a different category from the barges of the gentry.

Clearly then, George Crowninshield's famous *Cleopatra's Barge* of 1816, sometimes described as the first American yacht, was certainly not so, but yacht she assuredly was in conception and use. She was a brigantine of 83 ft on the waterline, 23 ft beam and with a registered tonnage, as measured at that period, of rather more than 190 tons. The elaboration of her furnishing, and such details as the glassware and silver service produced specially for the yacht, indicate how fully the conception of a yacht was realized in this vessel. But that she was something of a wonder of her day is indicated by the number of casual visitors who came on board – 900 in one day in Salem and 8,000 during the period of her stay later in Barcelona. She would not have caused such a sensation in Britain even in 1816. But she would have been unusual in having been built initially and fitted out as a yacht and her cost of 50,000 of 1816's dollars would have been impressive anywhere.

In the following year she made a cruise in Europe, sailing to the Mediterranean via the Azores and Madeira. That Mediterranean summer cruise, with its lavish entertaining, prefigures all the glittering display of rich cosmopolitan yachting, mainly in steam and later motor yachts, which subsequently became the most prominent feature of the French and Italian Rivieras, and persists today. Of this *Cleopatra's Barge*, so aptly named for the role, was the pioneer.

In another respect she was typical of so many early yachts. When her owner died suddenly, shortly after returning to Salem, the yacht was sold and converted into a working vessel. In the early days, in America as in Europe, the smaller commercial and naval vessel and the yacht were closely related; the one could quite readily become the other.

It was of particular advantage to American yachting that the working craft of the New World, as evolved during the Colonial period, had qualities more suited to the yachting role than those of the Old. As Herbert Stone, a founder member and early Commodore of the Cruising Club of America has written: 'But while we did not have many yachts, the merchant fleet of the United States was in the heyday of its existence, and the pilot fleet of New York harbour had to keep pace with the growing business of the port.'

When the *America* came to England in 1851 the New York Yacht Club was only seven years old and the *America* herself was essentially a working pilot boat; while in Britain yachts had been built as such for decades. But the American tradition of building working craft of exceptional functional efficiency led them in a single bound to the lead. American yachting from the first days to the present can be understood fully only in the light of earlier American developments in sail.

Decades before American yachting was on its feet, what became known as the 'sharp' model of hull was evolved at some time in the first half of the eighteenth century; a fine-ended type of hull whose narrow forebody ran into a midship section with steeply rising floors and high, hard bilges. It was a type removed in shape from anything evolved among European sailing craft

The impetus behind this striking development was the pursuit of speed. In the craft of Europe carrying capacity and staunchness were the main characteristics sought. Even when speed was a paramont requirement, as in the Revenue cutters, that robust fullness in line still prevailed, and was copied in the English yachts. The Americans had a faster model to follow for their early yachts in the sharp hulls evolved for working craft.

Several practical reasons justified the American pursuit of speed in their smaller fore and aft rigged vessels. But speed at this time was sometimes followed beyond the limits of economic or military justification. It became, in common with many other manifestations of American life at the time, a sign of the forces being generated among a people drawn from many lands, who buoyantly and looking only ahead were making themselves into a nation. A new great nation caught the spirit of speed at sea, caught it before others and were led by it in their ideas about one of man's oldest techniques – harnessing the wind to drive ships. Clearly this flair, ceasing early in the history of yachting to be of value in commercial and naval affairs, remained in the growing world of pleasure sailing

Opposite

Reconstruction in the Peabody Museum, Salem, of the main cabin in *Cleopatra's Barge*.

one that could raise spectacular results. This was shown continuously in the international yacht racing of succeeding years. It was to be demonstrated most publicly in the series of America's Cup contests.

Whether or not yachts should be equipped with a centreboard was one of the most fruitful sources of acrimonious debate between the two sides during the earliest contests for the America's Cup. The dominant place of the centreboard in American yachting, in contrast with the active objection to it, including rules against it for some years in British yachting, was a direct result of earlier seagoing traditions.

The precise origin of the centreboard or lifting keel, perhaps most picturesquely described as a centreline leeboard, is uncertain. That it was used by the South American Indians in prehistoric times now seems probable. In the early Middle Ages the Polynesians used daggar plates in their balsa-wood rafts. But until the eighteenth century the invention was lost. A sort of sliding keel was devised by the British naval officer, John Schank, while stationed at Boston, Mass., but it had no influence. In Europe, where the leeboard held such a firm place for so long, including in the Dutch yachts, the centreboard was not considered as an alternative. Experiments were made, as noted above, at the end of the eighteenth century by the Commodore of the Cumberland Sailing Society on the Thames but the device created no interest. In 1811 the Attorney-General of the U.S.A. granted a patent for fourteen years to Jacobs, Henry and Joshua Swain of Cape May, N.J., who, it was said, '. . . have invented a new and useful improvement on the leeboard, which they state has not been known or used before their application. . . .'

We may doubt the latter claim; but not the fact that it was in American waters that the centreboard first became common, and that it was in use at the end of the eighteenth century. It was adopted in pretty well all varieties of American fishing and trading craft confined to shallow waters like Chesapeake Bay, Albemarle Sound, and the numerous inlets up and down the coast. American yachting had its origin in shoal waters of New York Bay, Hoboken, South Brooklyn, Long Island Sound, where the convenient anchorages were shallow; an American yachtsman had the example, which the British had not, of numerous centreboarded working boats to guide as model or to convert.

Hence emerged two traditions in yacht design and yachtsmen's prejudices: broad and shallow opposed to deep and narrow, centreboards opposed to deep fixed keels – this expresses briefly the essence of American and British inherited ideas in yacht design, which initially reacted strongly on one another in competition often most acrimonious.

The keen American instinct for speed led, apart from the sharp hull, to developments in rigs, rigging and sails the merits of which were not evident to Old World yachtsmen until the *America* startled the Solent in 1851.

The origin of the schooner rig is obscure, but it is undoubted that its development was mainly American. In the U.S.A. it became the traditional national rig in the course of the eighteenth century, and Europe had to borrow the ideas that had produced the refinements of the rig. The American schooner was evolved to a high state before there were any yachts to carry it; once there where it became the usual rig of the larger American yachts until the yawl and ketch gained the greater favour in the present century.

At the time of the 1812 war, American and British navy alike remained faithful to flax canvas. But already privateers and blockade runners using cotton canvas were proving themselves to be more weatherly than the naval vessels. Initially woven on hand looms, the later use of power looms enabled cotton

The *America* in English waters in 1851, when still under American ownership. Painting by Fitz Hugh Lane.

cloth to be made harder and less stretchable. Until the twentieth-century invention of man-made fibres and hence of terylene and its like, cotton duck was the yet attainable ideal in sailcloth for the fore and aft rig. But when the *America* raced in the Solent, it was against a fleet of British yachts all canvassed in the soft yielding flax which had been the more suitable for square-rigged ships. And even then it has to be confessed that the British were not quick to adopt cotton.

Cleopatra's Barge may retain her honourable place as an early example of an American pure yacht, and the first perhaps to be essentially unlike any working boat; but the dawn of yachting in America is undoubtedly more clearly marked by the appearance in 1840 of the schooner *Onkahye*, though she was fundamentally of working-boat type, and indeed finished her days in the Revenue Service. Schooner rigged with cotton sails, she was in the best American tradition; but more remarkable at the time was her fine bow, the further development of which was to emerge ten years later in the *America*. But a technical ingenuity in the yacht, which anticipated the engineering expertise of gear and equipment, in which America was to lead the world of yachting in subsequent generations, lay in the fitting of sail tracks and slides. These did not become common until the adoption of the Bermudian rig in the twentieth century.

While *Onkahye* was in advance of her time, the sloop *Maria* launched in 1848, and like the former an inspiration of one of the mechanically minded Stevens family, was an astounding anticipation of future yachting techniques. She

Maria, the sloop launched in 1848, which with a hollow mast carrying track and slides, outside ballast, cross-cut sails and many other mechanical ingenuities, was a remarkable anticipation of the future.

John Cox Stevens, a founder member of the New York Yacht Club in 1844 and its commodore from 1844 to 1855. He had the brilliant mechanical gifts of his family, demonstrated in the *Maria* (see p. 38) which was named after his wife and of which he was part owner. He was later head of the syndicate owning the schooner *America*.

followed *Onkahye* in having tracks and slides. She had too what is believed to be the first suit of cross-cut sails, which did not become common until more than half a century later. Outside lead ballast was fastened to the hull in strips and covered with copper sheathing. Finally, she had a hollow mast. This 92-ft long mast, 2 ft 8 in in maximum diameter, was bored out to produce a wall thickness of 10 in for the first 20 ft above the deck. Above this the diameter of the boring was decreased to 10 in for the next 20 ft and for the rest of the length to 7 in. The boom was hollow also, constructed like a barrel of oak staves, dowelled together and hooped.

Four years after she was launched *Maria* had her bow lengthened by 18 ft, which made her 110 ft long on deck and 108 ft on the waterline. A 24-ft long heavily weighted centreboard dropped to 20 ft and was raised by means of spiral springs. While not longer overall than the biggest of subsequent racing cutters, on the waterline she was the longest single-masted vessel ever built. This yacht of 1844 anticipated the genius of American yachting far more vividly than the more famous *America* of 1851. She amazingly revealed the features of the great racing cutters which were to emerge under the stress of millionaires' and America's Cup racing in the 1890s and early years of the twentieth century. Over-bred, over-indulged, over-canvassed, mechanically ingenious, essentially fragile, absurdly expensive – *Maria* was the shape of all these things to come. And she was the fastest yacht afloat in the right conditions. But she was a sailing machine and unadaptable, though believed to have cost 100,000 of 1844's dollars. Dismasting occurred several times; while at her best in smooth water, and far faster than the *America* under these conditions, in heavy weather the latter and her kind were greatly her superior.

It was in 1935 that a brother of Starling Burgess, who was internationally famous for the design of America's Cup defenders, described such craft thus to the Society of Naval Architects and Marine Engineers:

'. . . the modern America's Cup racer bears not the slightest resemblance to any useful craft in the world, and she does not even contribute to the development of yachting as a true sport apart from the satisfaction of an illogical national vanity. But having damned them, I must confess to an absorbing interest in the problems set by those extraordinary craft. They have the fascination of sin.'

They were the product of over-indulged yachting sophistication, a type of craft never previously known in the history of sail, thrown up for a brief period by a tiny, but spectacular, section of the yachting world, now no more. *Maria* foreshadowed this fascinating, brilliant, rather decadent and spectacular kind of yacht and yachting.

The astounding thing is that *Maria* was being designed only three years after the New York Club was founded. Thus early, and in such a thin yachting scene, did the American yachting genius lay its future course. And it was on board a yacht belonging to John C. Stevens, one of *Maria*'s owners and brother of her designer Robert L. Stevens, that the club was established. In 1844 John C. Stevens was sailing a new schooner just designed for him by George Steers, who a few years later was to design the *America*. This schooner, *Gimcrack*, about 51 ft overall and 13 ft 6 in beam, was not herself a notable vessel and served Stevens for only three years, and a few years later was broken up at Oyster Bay, Long Island, leaving no record of her short life either in model or drawing. In her small cabin, while the yacht was lying off the Battery. New York Harbour, the New York Yacht Club was founded by nine men, eight of them yacht owners, late in the afternoon of 30 July 1844. John C. Stevens was elected Commodore, and remained so until his death in 1854.

In 1844 New Yorkers used to go picnicking by the flat shore known as the Elysian Fields, and here the club established itself in a modest building that has been more charmingly described as 'a handsome Gothic cottage in a pleasant grove'; it was on the property of the Commodore. By 1868 the membership was twenty-two and there were a dozen yachts in the fleet. In 1868 it moved to Clifton, Staten Island, and here was begun the famous model collection. Three years later the club also took city quarters in a house in Madison Avenue, 27th

The present home of the New York Yacht Club on West 44th Street into which it moved in 1901, shown in a pen drawing based on a photograph. It has been described as 'the finest building of the kind in the world', and certainly today houses the greatest collection of yacht models ever assembled.

Street. This indicated the club's growing social stature but not evidently its financial stability; for by 1876 it was economically in difficulties. The Staten Island station was closed, but a city premises retained. Eight years later this was moved to 27 Madison Avenue. It was not until the present century, in January 1901, that the club took occupation of what is and remains the most splendid building of any yacht club, near 5th Avenue in West 44th Street, where the present house was built on land given to the club by the financier J. P. Morgan. It was this member of the Morgan family who is believed to have made the remark 'If you have to ask what it costs to run a yacht, you can't afford one'.

Here now at 37–41 West 44th Street is the most entrancing display of yachting history to be found anywhere in the world, filling the famous model room and library, and here the present author has spent engrossed hours. There is no collection of yacht models to parallel, even to approach, that found in the innumerable glass cases and hanging on the covered walls; and here too, of course, is a model of the *America* herself.

At the time when the New York Yacht Club was founded there were already in existence a few other American clubs, none of which have survived; while in the British Isles, including Ireland, we have seen that there were eleven 'royal' clubs in 1844 – though not all had yet received this designation – which survive today, and the first British Commonwealth club had been founded – the Royal Gibraltar – as early as 1829. In the same year as the New York Yacht Club's foundation there was formed the Royal Bermuda and in Britain the Royal Mersey. The Royal Canadian Yacht Club, which was later to issue the third

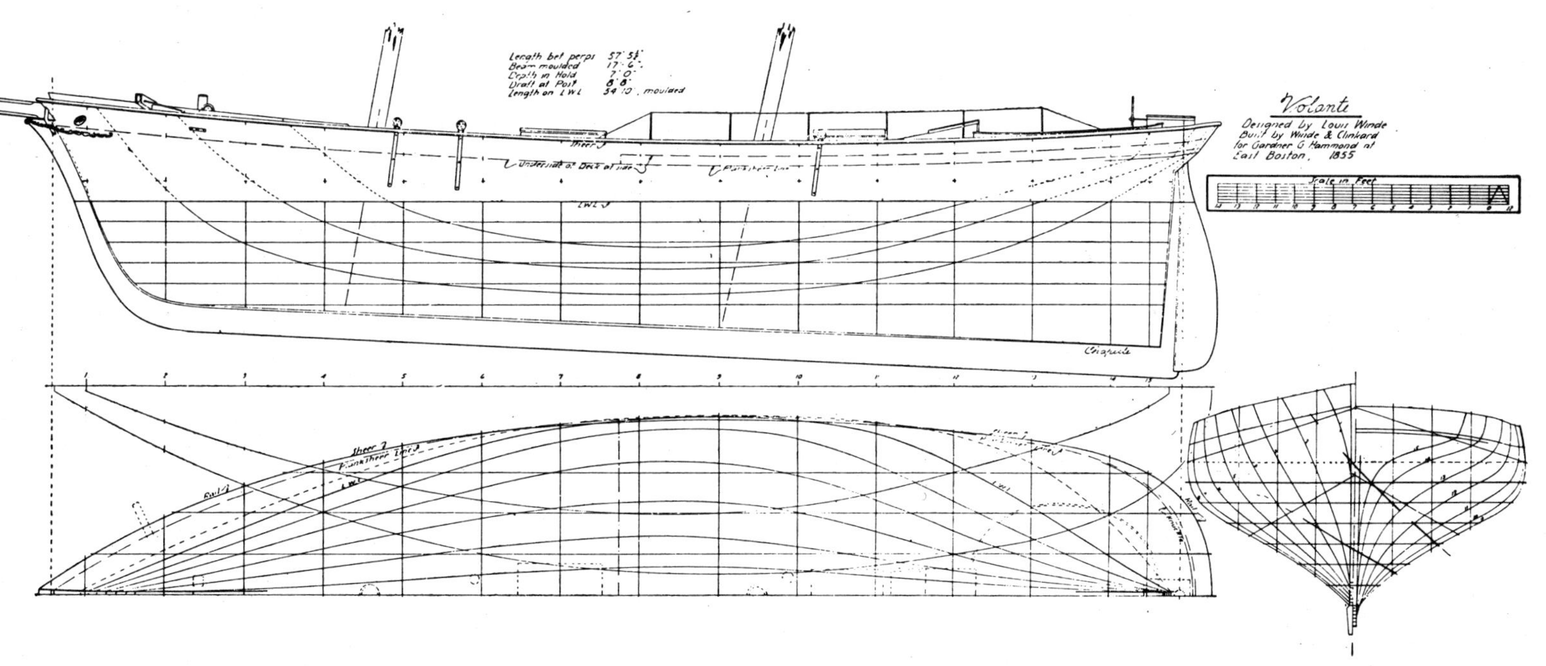

American yacht *Volante* of 1855 for comparison with British yacht *Volante* of 1851 (see p. 44). The American *Volante*, designed by Louis Winde, typifies a good American yacht of the period before extreme proportions had been adopted. Her length on waterline was 54 ft 10 in, beam 17 ft 6 in, draught 8 ft 8 in. Note the low freeboard, slight sheer and short ends, and compare dimensions with the English *Volante*.

and fourth challenges for the America's Cup, was founded in 1852, the year following the *America*'s visit to England.

At this time the New and the Old Worlds were about to react upon one another in the sphere of yachting, where hitherto they had not come within touching distance.

The two worlds shared little in ideas or prejudices, and were divided rather than united in a more or less common language. Thomas W. Lawson, an American historian of the America's Cup, writing in 1902 said '. . . there was very little social intercourse between the two countries . . . while there was a very low opinion in England of American social life'.

Whatever may have been their social prejudices, English yachtsmen were certainly blind to the technological achievements exhibited in the still comparatively small American yachting fleet. That the development of yachting in the U.S.A. had been to a large extent due to men with an expert professional knowledge of the sea; that the design and construction of American yachts often, though not invariably, showed the expertise of the age that was coming to birth, in contrast with the Old World's yachts rich in the prejudices of the one that was dying; that the Americans pursued competitive sailing with a ruthless professionalism foreign to the older tradition of the amateur along the coasts off which Nelson's ships had commanded the seas, was not recognized in the clubs where the load of tradition was heavy.

In yachting the Old World still had something to teach the New; but in the 1850s what the New had to teach was of more immediate and radical importance. The Solent needed a breeze from Sandy Hook. And this it received.

The *America*
Where did she come from? New York Town!
Who was her skipper? Old Dick Brown!

as the contemporary New York waterside song went – appeared through the evening gloom off Cowes on 23 August 1851, to cross the winning line for the Royal Yacht Squadron's Cup. Queen Victoria had been passed and saluted by the *America* as the royal yacht lay in Alum Bay and the *America*, in the lead in fitful light airs, was on the last leg of the course from the Needles to Cowes. Later the Queen was misinformed when told 'There is no second, Ma'am'.

Some 21–3 minutes after the *America* (records of the times differ) her closest rival, the little cutter *Aurora*, followed her over the line. *Aurora* was 47 tons compared with the *America*'s 140 tons, and but for the fact that time allowance had been waived for the race she would have been the winner by a handsome margin. She might even have won on elapsed time, for owing to muddled race instructions the *Aurora*, and several other English yachts, sailed a longer course. This most famous yacht race ever to have been sailed was, as a sailing contest, unsatisfactory. When the band on shore played 'Yankee Doodle' as the American schooner crossed the line, it was applauding unquestionably the best yacht amongst the nine schooners and nine cutters that had raced that day round the island; but under the prevailing conditions it was chance, accident and bad organization that produced so appropriate a result. Luck converted the Royal Yacht Squadron Cup into the best known of all yachting trophies, the America's Cup. But nineteen years were to pass before the cup re-enters history and many more before it became famous.

The course round the Isle of Wight, so popular in 1851, and becoming increasingly so ever since, is not a satisfactory one, with the strong tidal influences involved, and in July and August the fluky winds liable to prevail. And when the tonnage of the yachts involved ranges, as they did in the America's Cup race, between 47 tons and 392 tons, the contest becomes a lottery.

We need not re-tell the story of the celebrated, though confused, race; but the qualities of the *America* should be sung. She came amongst the fleet of the then-leading yachting nation as a craft radically differing in hull form and rig from all generally accepted British ideas of good design. She had no sooner reached the Solent than she showed her broad heavy counter to the English cutter *Laverock*, which had sailed out from Cowes to meet her. The short, unofficial race of about six miles back to Cowes resulted in the schooner coming to anchor some quarter or third of a mile ahead of the cutter. The 'glorified pilot-boat' as one English writer had described her, had walked away from a yacht that was believed by good English judges to be an adequate representative of the finest yachting fleet in the world. Englishmen began looking at the Yankee schooner with an interest sharpened by apprehension.

'A big-boned skeleton she might be called', one of them wrote. 'Hers are not the tall, delicate, graceful spars with cobweb tracery of cordage . . . but hardy stocks, prepared for work and up to anything that can be put upon them. Her hull is very low; her breadth of beam considerable, and the draught of water peculiar – 6 ft forward and 11 ft aft. Her ballast is stowed in her sides about her waterlines, and as she is said to be nevertheless deficient in headroom between decks her form below the waterline must be rather curious. She carried no foretopmast, being apparently determined to do all her work with large sheets.'*

Another writer wrote on 17 August in *Bell's Life* six days before the great race: 'When close to her you see that her bow is as sharp as a knife-blade, scooped away as it were outwards till it swells towards the stern. . . . Her stern is remarkably broad, wide and full. . . . Standing at the stern and looking forward the deck is nearly of a wedge shape, the bow being as sharp as the apex of a triangle, and the stern not being very much less than the extreme breadth of beam.'

These English writers had looked with some discernment on the craft that was so strange to their inherited ideas of maritime rightness. It will be evident that it was the hull shape that particularly intrigued them, the long fine narrow

Model of the *America* in the Science Museum, London. A number of contemporary half-models of the yacht were made, one of which is now in the Mariners' Museum at Newport News, Virginia. Various sets of lines were lifted from the yacht or models at different times, including one secretly and at night when the *America* was in an English yard in 1859. A finely rigged model, without sails, is in the New York Yacht Club.

* The use of the term 'sheet' to designate 'sail' is a nautical solecism.

entry which was in so sharp a contrast to the relatively bluff bows of contemporary English yachts. There was, however, one yacht in the English fleet at that time, designed by the eminent naval architect John Scott Russell – who later was to create the hull of the *Great Eastern* – which had a fine bow resembling the *America*'s. But this yacht, *Titania*, which in this respect was superior in conception to other English yachts of the period, had other marked inferiorities compared with the *America*, which resoundingly defeated *Titania* in a match sailed a few days after the Royal Yacht Squadron race. Scott Russell confessed the inferiority of his yacht with engaging frankness in his massive work *The Modern System of Naval Architecture*: 'On the wind *America* stood up whereas *Titania* heeled . . . *America* weathered *Titania* on every tack.'

Titania's lack of stability was due to the yacht's lack of beam, which was the outcome of the British tradition in design of narrowness but the fact that *America* was so much closer winded than *Titania* – and indeed than all the other British yachts – was due also to a feature of the yacht that the British were not quick to appreciate. As we see above, they observed the clean simplicity of the schooner rig, the simple sail plan, the lack of foretopmast, the single large headsail which she set; and they noticed too that the mainsail was laced to the boom, in contrast to the British yachts which carried their sails on their booms loose footed. They were less quick to notice the vital fact that the *America*'s sails were of machine-spun cotton. British sails up to this time were of hand-made loose-textured flax. That they became baggy when set was recognized as a fault, and attempts were made to remedy this by the process known as 'skeating' – the shrinking of the luffs of the sails by drenching them with water. The result remained much inferior to firm cotton sails.

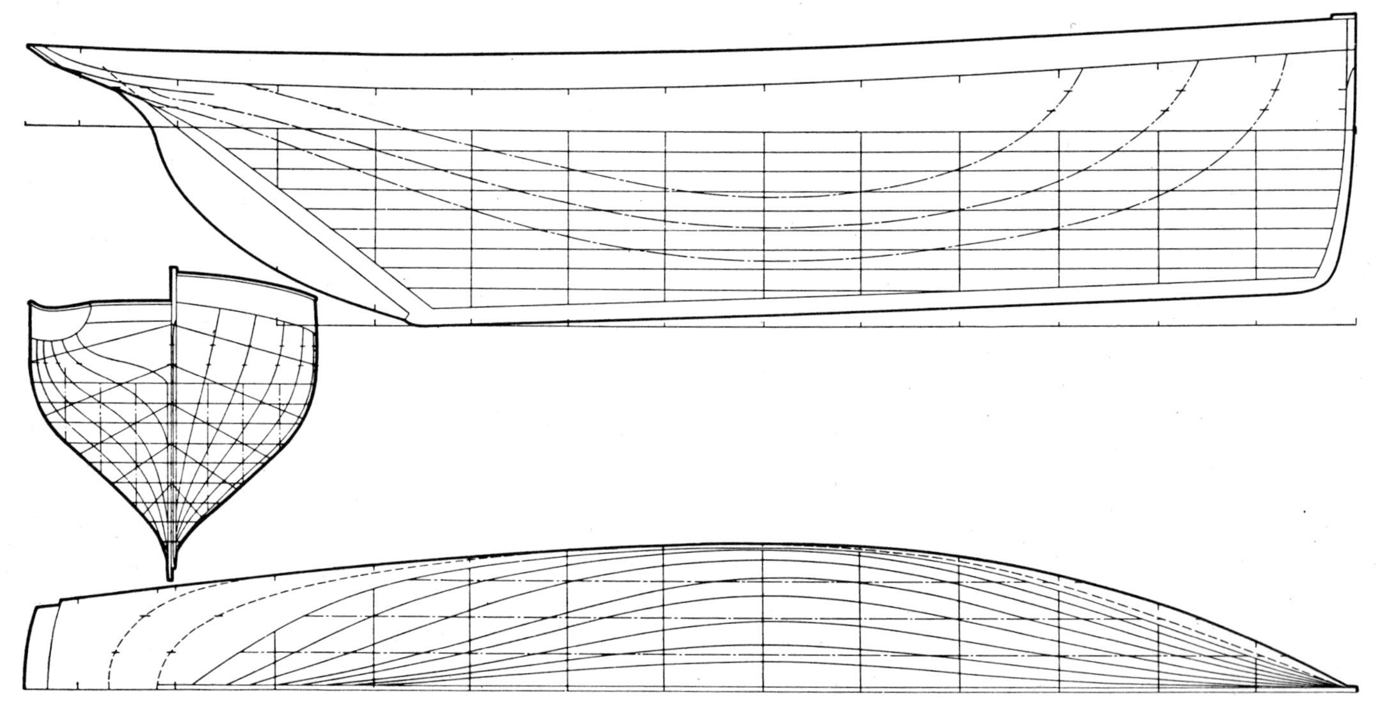

Volante of 1851, designed by John Harvey, a member of the Institution of Naval Architects, London, and built at Wivenhoe, Essex, England. Length waterline 61 ft 6 in, beam 14 ft 9 in, draught 10 ft. The extreme rake of sternpost was to reduce the tonnage measurement. In her proportions she typifies the good English yacht of the period before extremes were reached.

It was a century after the *America* had sailed her race that an American yachtsman still felt justified in saying 'The British have traditionally never been able to understand sails,' and the stricture no doubt continued to be a reasonable one. It is not to be questioned that of all the *America*'s superiorities, in hull form, in rigging and basic sail plan, it was the stuff of which her sails were made, their flat cut and the lacing of the mainsail to the boom, that gave her most advantage. From the example of the *America* British yachtsmen learned the advantage of cotton duck sails; but it is an interesting manifestation of national traits that persist through generations, that in the years of international yachting competition between the U.K. and the U.S.A. the latter has persistently led the way in the design, cutting and setting of sails. This is a theme running through the international yacht-racing story, though at some times more dominant than at others. It should perhaps be added that when the *America* passed into British ownership the lovely cleanness of her rig was spoiled. She was given a foretopmast; the great forestaysail, reaching out to the bowsprit end, was discarded (*America* carried a jib on a jib-boom for part of the famous race, until she carried it away) and ahead of the mast she set the English rig of small staysail, jib and jib topsail instead of the pilot boat rig with which she arrived in English waters.

The *America* gave the most convincing demonstration of her superiority to English yachts – 'If she's right we must all be wrong' the eighty-year-old Marquis of Anglesea had observed – in a race for which she was not officially entered.

It is often incorrectly said that the cup she won at Cowes was the Queen's Cup. It was simply a Royal Yacht Squadron Cup, and the Queen's Cup was sailed for two days later. The wind being light that morning the *America* did not start; but when it freshened she sailed an hour and a half later. The following account of her performance appeared in the *Illustrated London News*:

'Just before the vessels got in, the raking American was seen making her way round the Nab Light, and, with a most extraordinary movement, made one reach from the light to Stokes Bay, and by another tack, rounded the *Brilliant* in gallant style. To accomplish the same feat that the *America* had performed, the *Alarm* took ten tacks, and the *Volante* at least twenty for the same distance.'

After her late and unofficial start the *America* is reported to have come in a few minutes after the winning yacht.

The Royal Yacht Squadron's 100 guineas cup was carried to the U.S.A., though not by the *America*, which had been sold in England. It was not until nineteen years later that it returned to the stage of international racing, in 1870, by which time the *America* was back in the U.S.A., owned now by the navy; and she took part in the 1870 defence.

For six years the cup provided heavy baroque decoration in a Washington Square drawing-room, that of John Stevens, first Commodore of the New York Yacht Club. It was put back into circulation when the surviving members of the *America*'s syndicate determined to offer it for international competition. The outcome was the cup being offered under a simple Deed of Gift, to the New York Yacht Club, now in its thirteenth year, for 'friendly competition between foreign countries'. The salient points of the Deed were that the trophy was to become a perpetual challenge cup, that challenges were to be made through an established club, not by an individual, and that the winning club was to hold the cup in trust. Terms of a challenge were to be agreed by mutual consent. It transpired that the latter was often difficult to achieve.

This was the summer of 1857. Twelve more summers were to pass before anything more was heard about the cup. There was no hint yet of the almost frenzied interest to which it was to give rise, of the wealth to be expended for its possession. It later gained a status not only in yachting, but overspilling into social history, that deserves to be rationally analysed on both scores.

To the man in the street it became, and to a lesser extent remains, the supreme event yacht racing has to offer. During a long period it was the one aspect of yachting with which he was even slightly familiar. In years when the contest was being held 'cup talk' would be heard far from the waterfronts, wonderful solecisms would be written in the newspapers published in localities where maritime news was rarely considered worth attention; while in Britain America's Cup news would appear on the front pages of the national newspapers instead of inside on the sports pages. International fame came to those involved in the challenge and defence operations, which assumed the likeness of war without weapons; and sometimes the disputes that the contests set alight made a parody of the 'friendly competition' for which the cup had been presented, even becoming of diplomatic concern. It has been said that one British ambassador in Washington declared with exasperation 'Damn the America's Cup'. Closer to the sea there was, even for the first contest in 1870, a vast fleet of sightseeing craft, a Venetian regatta in a New World setting, and the Wall Street exchange was closed for the day.

It was this frenetic sense of association with the cup felt by so many landsmen that gives it social significance. But the contests produced something of greater importance to yachting itself than this circus appeal. While modern yachting was still young the America's Cup brought together two opposed types of yacht and yachting society. During the early years of cup racing the yachts involved were not extremely specialized types of craft, but generally representative of the largest and best racing yachts of Britain and the U.S.A. This was so during the seven contests that took place between 1870 and 1887. These were crucial years in the evolution of the modern yacht, when two opposed traditions in design were in strenuous competition. In days when there was little other international yachting competition, and yachts of some size were representative of the most powerful section of the sport, America's Cup racing became a forcing house of design development and an influence upon yacht types in all sizes. The long shadow of that development reaches to our own time.

After 1893, when the eighth contest took place, America's Cup yachting drew progressively farther away from the main stream of yachting. The yachts themselves became of an esoteric type, specialized to a degree rising occasionally to a fine frenzy of eccentricity making them unlike any other yachts anywhere. But thanks to their swaggering extravagance, and the expertise to which their handling raised the art of racing, they retained a firm hold upon the nautical imagination, while entrancing the millions who watched from afar the millionaires' amazing sailing performances.

The first challenge for the cup introduced immediately, and in the slightly acrimonious form that anticipated the future in other ways, the conflict of types between American and British yachts, both of which were to become so much altered as a result of later contests. It has been seen in previous pages that in the American tradition of seafaring the centreboard, or sliding keel, held an important place in small craft, and as a result centreboard yachts with broad-beamed, shallow hulls became a dominant type. In European yachting, on the contrary, centreboards were regarded as unseamanlike devices not suitable for the best types of yacht. A rule of the Royal Yacht Squadron was uncompromising in the matter: 'No vessels which are fitted with machinery for shifting keels, or otherwise altering the form of their bottoms, shall be permitted to enter for races given by the Royal Yacht Squadron.'

In 1868 an American schooner aspired, and failed, to emulate the *America*'s performance in English waters. The schooner *Sappho*, length on deck 133 ft 9 in and on the waterline 120 ft, without centreboard and with a draught of 12 ft 6 in, was the largest yacht yet built in the U.S.A. Racing in England with the rig under which she had crossed the Atlantic and carrying the stone ballast shipped for this voyage, she was not in good trim for racing. Amongst the

British yachts that defeated her round the same course as that sailed by the *America* was James Ashbury's schooner *Cambria*. With this yacht Mr Ashbury proposed not so much a challenge for the America's Cup but a series of races with an American schooner of approximately the same size as his own, including a race from east to west across the Atlantic. Ashbury was one kind of typical Victorian, self-made and newly rich through industry and carrying his native aggressiveness into sport. He found a match, however, in the New York Yacht Club, now custodians of the trophy. During these early years the preliminaries to cup contests involved negotiations resembling commercial treaties by tight-

On board the first challenger for the America's Cup in 1870, the schooner *Cambria*, as illustrated in *The Graphic* for 28 May of that year. A tackle has been put on the tiller to ease the weight, while members of the crew and afterguard sit up to weather. The sartorial styles of the latter were less functional than than those of today.

fisted business men with eyes glued to the main chance. Both sides disregarded the Deed of Gift, Ashbury by trying to impose quite different terms of his own, and the New York Yacht Club terms of marked inequality favouring the Club. The wrangling which began in 1868 persisted until the first America's Cup race was sailed in August 1870. Ashbury had wished to race against a single yacht only and had originally insisted that he would not meet a centreboarder. He ended by having to race against a fleet of twenty-three yachts, sixteen of which were centreboarders.

The race itself had a more impressive prelude in the match sailed across the Atlantic by the *Cambria* and the *Dauntless*. The latter was a crack American ocean-racing schooner, and she carried on board for the transatlantic match the 'Ole Dick Brown' of the *America* in 1851. The fact that *Cambria* won the race of some 3,000 miles from Daunt's Rock to Sandy Hook, though only by 1 hour 43 minutes, was disturbing to American opinion.

The America's Cup race itself lacked interest. Twenty-three yachts determined to prevent the twenty-fourth winning were conditions not comparable with those faced by the *America* in 1851, when she encountered not a fleet united in the object to stop her winning, but one in which every yacht sought to win herself. The race, which was sailed with time allowance, was won by the gracious schooner *Magic*, a centreboarder and one of the smallest of the fleet, which had been built originally as a sloop.

Records indicate that during the remainder of that summer, which James Ashbury spent racing *Cambria* in American waters, the social climate was amicable, though there was little racing success for the British yacht. This was significantly attributed to her sails and rigging rather than hull. But when Ashbury, after returning to England, determined to challenge again in the following year with a new yacht, the wrangling resumed, rising to its most acrid when the America's Cup Committee reminded Mr Ashbury of 'the fact that the deed of gift of the cup carefully guards against any sharp practices'. The sharp practice complained of was James Ashbury's proposal – the challenge being between clubs not individuals – that he would issue twelve challenges, one on behalf of each club of which he was a member (one of these was the Royal Albert Yacht Club, where these lines are being written) and demand the sailing of twelve races, victory going to the first to win seven races.

Ashbury, in fact, was proving himself as hard a bargainer as the New York Yacht Club, which had already, at the instigation of the surviving member of the syndicate that had presented the cup, agreed that a challenger should race against only one yacht at a time. This was a ray of light crossing the murky scene where possible sailing rivals manœuvred for position in telegrams equivocal, ambiguous or rude, before taking to the water. These hardy mid-nineteenth century business and commercial men, unaccustomed to let advantages slip, were by trial and error and many false starts, grinding out an acceptable ethos for fair yacht racing.

We may recognize this fact and not go further into the acrimonious controversies that preceded the next contest. Ultimately Ashbury accepted that he must face a centreboard yacht or one of deep keel, according to the defender's choice; and furthermore, the Cup Committee had interpreted the ruling of the cup's donor that the challenger should have to meet only one defender as meaning 'only one defender at a time'. The contest was to be the best out of seven races, and the defence reserved the right to nominate any one of four yachts, the keel schooners *Dauntless* and *Sappho*, mentioned already, and the centreboard schooners *Columbia* and *Palmer*, for the race on any particular day. The former pair were at their best in heavy weather, the two latter in light. The defence might choose the yacht to suit the weather. The conditions were not much better than in the former contest.

If the Cup Committee could be difficult – and some Americans considered their quibbling attitude unsporting even in 1871 – James Ashbury would match them step by step. After a severe Atlantic crossing *Livonia* came to the line for the first race in a light wind on 16 October. The centreboarder *Columbia* was chosen to defend. With her centreboard up the defender ran away at the start from the challenger in the light reaching wind, the challenger dragging her 12 ft 6 in of draught through the water against *Columbia*'s 5 ft with the board raised. In the light weather the *Columbia*'s lead remained throughout the course, and she won by a big margin.

The Americans made the mistake of again selecting *Columbia* to defend in the next race, when the wind was brisk at the start and freshened later; but they were saved from the results of their misjudgement by a lack of clarity in the race instructions which favoured the defender. It was a poor course, to an outer mark and return with no windward work on either leg. Instructions were not given as to which side the outer mark should be rounded. The American skipper queried this with the Committee and was told it might be rounded on either side. The British assumed that, as in their own rules, when no instructions were given a mark should be left to starboard. This probably lost her the race. *Livonia* was leading at the outer mark, but leaving it to starboard forced her to make a fierce gybe all standing, setting her well down to leeward, while *Columbia* tacked round the mark under her stern, and from a position nicely up to weather led *Livonia* over the line. Mr Ashbury protested on several scores, but the protest was not upheld. A journal *The Spirit of the Times*, which the American historian of cup racing, Thomas W. Lawson, has described as the paper that 'voiced the best sentiments of the American press in yachting at that period' objected to this decision, as did several other yachting writers later. So was America's Cup history opening its story of protests and of race committee decisions often condemned by the Americans themselves.

Neither *Columbia* nor her crew were in a fit state to race next day, the crew after celebrating two victories and the yacht having a sprung foremast and with rigging that needed setting up. The challenger, of course, had equal reason to be suffering from rigging defects after two hard races. The organization of the defence, however, was not of the standard that became usual in later years. None of the other three specified yachts was in a condition to race also; and though it was suggested that *Magic*, winner of the last year's contest, should join the defence, and Ashbury was prepared to meet her, the Committee ruled it to be unfair that any other than the four reserved yachts should race. They were men trying to be fair, despite their dubious protest decision, under conditions of stress and fiery international yachting competition for which there was yet so little precedent.

The best defender for this third and heavy-weather race might have been *Sappho*, but for some curious reason she was in dock. Rather forlornly *Columbia* appeared late for the start with a crew of many substitutes, including one for the sailing master, and what was described after the race as 'too many amateurs'. She was over-canvassed; at one moment was in danger of capsizing with her shallow dish form. Under the pressure of canvas her wheel steering gear broke down, and hectic moments were wasted while axes were used to break open the steering box and enable a tiller to be rigged. Though steering badly under tiller she was coming up on *Livonia* when either the fore topmast stay or the staysail sheet parted, and the sail blew itself to ribbons. She shortly retired. The *Livonia* won the first of the only five individual races out of forty-nine ever won by a challenger in the sixteen contests raced until 1958.

The score in the contests was now, according to the race committee, two races to one in favour of the U.S.A. and according to *Livonia*'s owner, who considered the second and protested race to belong to his yacht, the same score in favour of

Below

Mischief (leading) and *Atalanta* running under spinnakers in the America's Cup races of 1881. The sails carried by the former should be noted: topsail with topsail yard and jackyard, balloon jib and spinnaker of the flat kind which remained in use until the invention of the parachute spinnaker in the twenties of the next century. The flat spinnaker began to replace the square sail after 1866.

Opposite

Irex, whose lines are shown in the top drawing, represents the typical English cutter of the 1880s. She was narrow, with between five and six beams in her waterline length. *Gracie*, in the lower drawing, is equally typical of the American – beamy, shallow, with centreboard. Her life spanned the transitionary period in American design, represented by *Volunteer*. She was extensively altered during her career, being increased in overall length from 60 ft 3 in to 80 ft, in beam from 18 ft 8 in to 22 ft 6 in, in draught from 5 ft to 6 ft 8 in, and finally she carried most of her ballast externally whereas originally it had been all inside. The changing *Gracie* embodied in herself the development of American yacht design during two decades.

Britain. He sailed the fourth and fifth races 'without prejudice' to the protest. Both these were sailed mainly in fresh winds. *Sappho* defended and won both. The score was now, according to Mr Ashbury, three to two against him, with the contest still to be decided by the next one or two races; the defenders considered they had already secured victory with four races won out of five.

There followed some *opera bouffe* in which *Livonia* claimed a sixth race because no officially named defender came to the line, though on this occasion one of the reserved four yachts, *Dauntless*, sailed a private match with *Livonia* and beat her. He also claimed the seventh race because no other yacht came to meet him. He did so without sailing round the course.

Two Canadian challenges for the America's Cup, in 1876 and 1881, followed the first two by Ashbury, and left little impression on yachting history, even of ill will; but they marked two not unimportant developments firstly in cup racing, secondly in racing-yacht design.

The first was that the New York Yacht Club now agreed, after some hesitation, that the whole defence should be conducted by one yacht only, thereby eliminating an obvious source of inequality in the competition. Next, in the second of the challenges, the Canadians produced a cutter (or American sloop) and thereafter two-masted vessels disappeared from cup racing, as they were doing so from inshore racing everywhere as the great day of the racing schooners went into decline.

The Canadian challenges were the only ones, until the eighteenth challenge in 1962, when the contest was other than between the U.S.A. and U.K. They

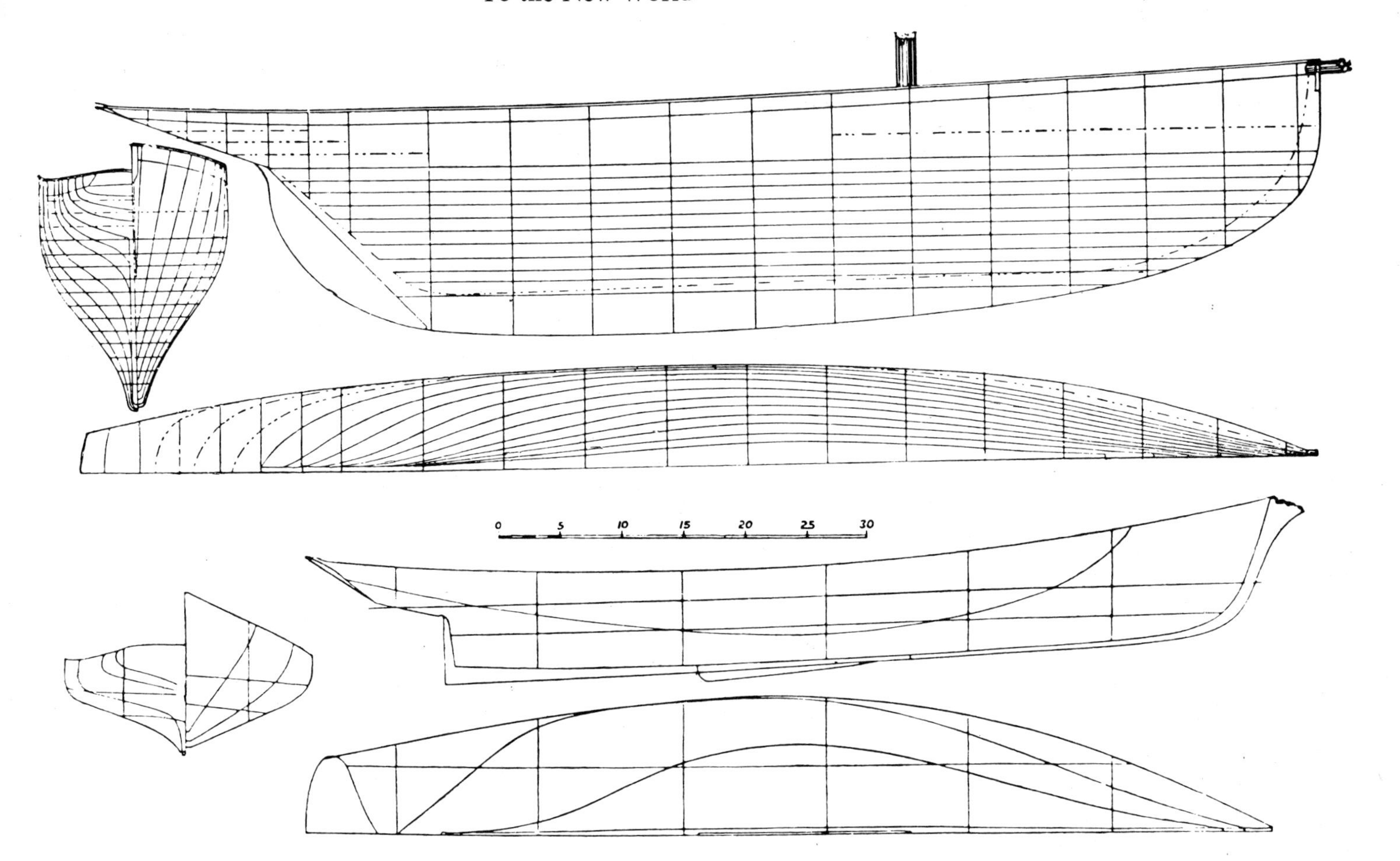

Volunteer, the seventh defender of the America's Cup, of Edward Burgess' compromise type but with the large centre-board. The model is in the Science Museum, London.

came from the Royal Canadian Yacht Club, founded in 1852; and on the Great Lakes, where many members sailed, there were centreboard craft in the New World tradition able to match the dominant type of American yacht with their like. Both challenges, however, were easily defeated, being by boats relatively poorly built and rigged, though not deficient in sound basic design. The initiative of the Canadians in producing a sloop for their second attempt forced the New York Yacht Club to follow suit. The day of the schooner for America's Cup racing was already done. But when late in the season of 1881 the Canadian sloop *Atalanta* met the American sloop *Mischief*, there was little doubt of the outcome. Lack of financial backing ruined the hopes of both Canadian challenges before they had been formulated. The challenger was described as having 'a bottom as rough as a board fence', while the defender, despite having an iron hull and unusual in this respect, had a bottom 'scraped, painted, varnished, and pot-leaded until it shone like a well polished boot'.

The two Canadian challenges disturbed the New York Yacht Club as bringing into disrepute the high standards of the yachts and sailing which the Cup – it was only just becoming known as the 'America's' instead of the 'Queen's' – had been intended to inspire. It was felt too that the cost and labour involved in the defence had been wasted. Many years later, when the British challenged in 1958 and 1964 and went down to resounding defeat on both occasions, the American sentiments of 1881 were again aroused, if less vocally.

It has been seen that among American working craft of sail the fairly beamy shallow type with centreboard was common, and that yachts tended to follow suit; though it is to be noted that extreme beam and lack of draught were not characteristic of early American yachts. For a number of years the keel held its own with the centreboard. The *America*, widely different in hull form though she was from the British yachts she sailed against, was without a centreboard. But gradually the yacht, following its own line of development, became more extreme in beam and shallowness. The trend was not general, but sufficiently so to become a bad habit.

During the 1860s, '70s and '80s, broad shallow sloops with centreboards were the dominant type of American yacht. Following the *America*, and in the decades before the 1880s, there was a decline in the technical skill of those who designed yachts. Those, such as George Steers, *America*'s designer, who produced the early yachts for the wealth and society of New York and Boston, amongst which were many with wide knowledge of ships and the sea, were men of some science and mathematical capability, who produced yachts from the drawn plan and made the relevant calculations from them. With yachting becoming more fashionable and yachtsmen less discerning, a relatively crude breed of boatbuilder came for a time to dominate the scene.

Howard I. Chapelle has gone so far as to write of this time: 'It can be said that the designing of yachts had passed from the shipbuilders to boatbuilders in the 1850s, and from boatbuilders to carpenters in the 1860s. In the 1870s saloon keepers, a hatter and a fisherman, without any training as boatbuilders or shipwrights, became the leading builders and modellers of yachts!'

The attainment of more speed was assumed to be a matter of adding to beam and reducing draught, thus retaining sail-carrying power while producing hulls which became ever-shallower dishes. The largest sloops, beneath their immense spreads of canvas, might have no more than sitting headroom below deck. While their stiffness, or resistance to heeling, was little short of tremendous at the lesser angles of heel, their range of stability – lacking draught and ballast as they did – was small; and the years were scattered with yachting tragedies which mistakenly became widely attributed not to defective design but to the inevitable fortunes of the sea. The *Eva* capsized and was lost on Charleston Bar in 1866; the larger schooner *Meteor* capsized three years later. The greatest of the disasters was that of the *Mohawk* in 1876, a schooner of 140 ft in length but drawing only 6 ft of water, which was lost when lying at anchor with her sails set. She was struck by a squall coming off the high hills of Staten Island, and capsized drowning her owner and his guests, who were below deck and pinned down by the furniture crashing leeward. In this yacht the base of the huge rig, from the end of the bowsprit with its flying jib-boom to the outer end of the great overhanging main boom, was 235 ft, nearly 100 feet longer than the hull. In 1883 the schooner *Grayling*, a centreboarder built for racing and drawing but 5 ft 9 in on a waterline length of 81 ft, capsized on her trial trip not far from her moorings when struck by a squall of no great severity.

But by this time the lethal character of what was regarded as the cream of American racing yachts was becoming appreciated. There were those who had condemned the *Grayling* as dangerous before she had so quickly and surely proved the fact. The time had come when America was to learn from Europe. A body of American yachtsmen, led by C. P. Kunhardt, whose brilliant writings secured his place in American yachting history, forced a change in the prevailing dangerous concept of fast yachts.

Opposite, above

Schooner *Cambria*, first challenger for the America's Cup, running past the Sandyhook light vessel *c.* 1870. Painting by C. R. Ricketts, 1872.

Opposite, below

New York Yacht Club regatta *c.* 1850. Painting by Fitz Hugh Lane.

SANDYHOOK

Part II

Nineteenth century growth 1850-1914

4 Yachting becomes established

Big Class yacht racing – by which we mean the racing of the largest type of pure-bred racing yacht round inshore courses – had one built-in defect that always made it an affair of ups and downs. Like the short inshore waves over which the yachts rode, crest was quickly followed by trough, leading to another crest. Thus it was from the beginning of such racing in the Victorian period until it came to the wave 'that never a swimmer shall cross or climb' and one of the most beautiful, if relatively fragile and useless, types of sailing craft disappeared for ever with the guns of 1939.

The glow that settles easily over the past may give the impression that during the grander ages of yachting the summer waters were invariably rich with magnificent and enormous yachts. Many large yachts there always were; but they were not always first-class racing yachts. The latter, even in the past unspecializing days, were esoteric vessels designed to increasingly intricate rules for the sole purpose of racing. Even larger, and very much more comfortable yachts were common under the summer suns of yesterday; but the great racing yachts whose names are remembered, and gripped the imagination of a wide, non-seafaring public in their day, belong to a select few. Not many owners had more than a superficial interest in seamanship; it was even considered in some quarters that to be too engrossed in such a manual activity was socially improper.

The Big Class at any time was just a handful of boats whose racing depended upon the enthusiasm of a few rich owners of individuality and often whimsicality. Their decided opinions were not invariably based on any profound knowledge, and prejudice was an inalienable right of free men. If two or three of them decided they did not like the organization of the sport, or fell out with each other, they decided to go steam yachting or fast motor boating, and the class would collapse. On the other hand, hardly more than the rumour of a couple of new boats being laid down might produce a revival in Big Class racing. The pleasure of racing with royalty might be the magnet. These great racing craft, tending to evolve into racing machines, were a product of Vanity Fair – perhaps at its best.

The vanity fair of yachting was at its most surpassingly brilliant during the period of great schooner racing, which was at its height in the 1860s and early 1870s. This American rig had, on both sides of the Atlantic, reached its apogee in the schooners of the yachting fleets. One of the most beautiful of all man-made things, it has been said, is a schooner-yacht under all plain sail. After the 1870s great schooners were still built; two of the largest and most magnificent, *Susanne* and *Westward*, were built in 1904 and 1910; while *Westward* was still out racing with the Big Class fleet in 1935; though her speed in strong reaching winds, so much greater than that of the modern cutters against which she then raced, could rarely be displayed against close-winded cutters whose rigs were unable to stand any considerable weight of wind.

During the 1860s the big racing schooners were the leading feature of the

Sappho beating *Livonia* in the America's Cup race off Sandy Hook on 21 October 1871.

regattas in which they appeared, and at this time there were up to thirty of them out regularly. In English waters one of the most celebrated was the *Egeria*, with a length of a little less than 100 ft. She came out in 1865, and in 1872 was still being regarded as the fastest schooner afloat. Writing of her some years after the decline of schooner racing Sir George Leach said: 'The successes of the *Egeria* led to her being classed as a sort of standard or test vessel, and, taken all round, she was probably the fastest schooner we had, although in strong winds she was often overpowered by her larger rivals. Year after year vessels were built to beat her, but, kept up as she was in the best racing condition and well sailed by her skipper, John Woods, she proved, even to the end of her racing days, no easy nut to crack.'

Sometimes American schooners were in British waters. The 310-ton *Sappho* paid a visit in 1868, and was handsomely beaten, unlike the *America*, in a race round the Isle of Wight. In the course of the race she broke her jib-boom, but the superiority of the British schooners, both with the wind free and close hauled was none the less clear. A book could worthily be written about the vintage days of schooner racing, which lasted so short a while though longer in the U.S.A. than Britain, its brilliance as a combination of sport and spectacle. There is space to describe one race only, in an account by Sir George Leach; it is the Prince of Wales Challenge Cup of 1874:

'The race had an interest of its own, from the fact that for the first time an American yacht, the *Enchantress*, 329 tons, the property of Mr Lubat, competed for it. Mr Fish, the well-known American yacht-builder, had been brought over

New York Yacht Club regatta in 1869, showing a start from the stake boat in the Narrows off Staten Island.

to superintend the preparation of the yacht for the contest. The entries were *Enchantress*, *Egeria*, and *Shark*. The Cowes week this year was characterized by blustering winds and rain. The *Egeria* had sailed for the Town Cup on the Wednesday, and as the weather was bad, and she had got everything soaked, the race was postponed by consent from Thursday until Friday, the 7th, to give her a chance of drying her sails. The morning broke with a strong S.W. wind, and as *Egeria*'s skipper knew well what he was to expect in the Channel, he reefed mainsail and bowsprit, and housed topmasts. Although the *Enchantress* had to allow the *Egeria* a lot of time, being more than double her tonnage, it was felt on board the latter that, if the *Enchantress* got round the Shambles first, she would in all probability reach clean away and save her time: therefore that *Egeria*'s chance was to beat her adversary in the turn to windward, and so if possible increase the time she would have to receive from the Shambles home. The race was to be started at 6 a.m., and *Egeria* was early under way, prepared, with such a wind, which kept increasing every minute, for a hard fight. To the great satisfaction of those on board *Egeria*, the *Enchantress* was observed soon after coming down with a cloud of canvas over her. "Hurrah!" was the word; "something must go before long."

'*Egeria* gained five minutes at the start, and as in coming round after the first board on the Calshot shore her opponent got in irons, she gained full another five minutes. They had not made many tacks before the man who was looking-out on *Egeria*'s lee side cried, "There goes *Enchantress*'s jibboom!" which made *Egeria*'s crew feel that, although the weather was getting more stormy and dull,

The 'wonderful *Egeria*' as she was called, a schooner which came out in 1865, when schooner racing was the apex of yachting. She remained prominent for more than thirty years, though outliving her days of glory.

their prospects had considerably brightened. Without her jib-boom *Enchantress* was no match for *Egeria* in the beat to windward, and when the latter was well outside the Needles, and had passed the Shingles Buoy, *Enchantress* had hardly reached Hurst Castle. Just before she got opposite Yarmouth, *Egeria*'s second jib was blown clean out of the bolt-rope. This will give some idea of the strength of the wind, which made the sea outside exceedingly heavy. *Egeria* was standing in for Christchurch Bay with the view of smoothing the water, when she observed that the *Enchantress* was put before the wind, and turned back for Cowes, having carried away her fore-stay. Her competitor being placed *hors de combat*, *Egeria* was immediately put under snug canvas, and sailed easily until she rounded the Shambles Lightship at 4 o'clock. The run and reach home to Cowes round the Island were comparatively easy sailing, and she showed her blue light passing the Squadron Castle at 11 hrs 50 mins, thus winning the cup for the third time.'

That was in 1874, when schooner racing was already in decline. Four years later in Britain it was in full eclipse. There were several reasons for this. One was the growing popularity of the steam yacht. The racing schooners had been no specialized racing shells, but robust and if necessary (which was not often) ocean-going vessels. They had comfortable accommodation in which an owner's family and friends might gather as in a country house. But the steam yacht offered more comfort and greater reliability on passages from port to port. Furthermore, as a racing yacht the cutter was swinging into its day of supremacy which was to continue until the racing of big yachts ceased for ever. We have seen that in 1876 the America's Cup was contested for the last time between schooners. Still in 1881 the best yachts in the New York Yacht Club were schooners; but for the crack racing their day was just about to pass; and so it was on both sides of the Atlantic. The day of the cutter had come.

They were great pure-bred racing yachts ranging between the 70-tonners and 40-tonners, designed to the Thames measurement rule which in these years

Cutter yachts off Liverpool in 1853. The square cut of the main topsail is to be noted as typical of the period. Paintings by Samuel Walters.

produced what became known on both sides of the Atlantic as the traditional British cutter. John Chamier has produced a nice description of them: 'Great black-hulled, sturdy boats they were in those days – and built like churches to the glory of God and to last for more than one generation. They had great ferocious bowsprits made from full grown trees. Their stems were hard, uncompromising,

Jullanar, a yacht representing a seamark in the progress of scientific yacht architecture. Designed by an amateur, E. H. Bentall, she was shaped to gain the maximum length under the measurement rule of 1875 – hence the rudder post placed far ahead of the waterline ending – while having also the minimum of wetted surface.

1830	Revenue Cutter	$2\frac{3}{4}$ Beams in length
1847	*Mosquito*	4 Beams in length
1873	*Vanessa*	$4\frac{3}{4}$ Beams in length
1875	*Jullanar*	6 Beams in length
1882	*Chittywee*	$6\frac{1}{4}$ Beams in length

upright pillars of oak. Their topsides were as thick as a man's arm and ringed around with bulwarks like battlements.'

Yachts, in fact, had not yet become over-emphatic in their inclination towards evolving into racing machines; but that day was at hand. Design was only just beginning to gather the refinements of science, but it was about to do so with a rush. The inherent faults of the Thames tonnage measurement rule were to be taken ruthless advantage of at the hands of yacht designers and builders, abetted by owners intent on winning, as the age of innocence in yacht racing gave way to the age of science. We must now consider some aspects of yacht racing in the 1870s and 1880s.

The fact that under the Thames measurement rule, the narrower a yacht was in proportion to her length the lower her tonnage would be for that length, at first produced no ill results on the breed of yacht produced, because a reasonable amount of beam was necessary to provide sail-carrying power. The various tonnage racing classes ranged in size through 3, 5, 10, 20, and 40-tonners up to the splendid 70-tonners. The ballasting of boats by means of an external lead keel, though tried as early as the 1840s, was distrusted. The boats had all, or nearly all, their ballast internally; never more than a tenth externally. In type they were the last direct descendants of the first British yacht, also '. . . built like churches to the glory of God'. But once the prejudice against outside ballast was abandoned – and rumours of *Maria* reaching across from Sandy Hook to the Solent may have helped towards this end – a radical transformation in yacht type might occur.

While the Thames measurement rule assumed that the depth of a yacht was equal to half the breadth, it did not enforce that it *must* be in this proportion. There was no limit to the depth and draught that a yacht might be given other than the powerful curb of earlier practice, so strong in maritime affairs, and the limits set by the constructional capabilities, which developments in engineering were extending. The draught of yachts was increased in proportion to their length, the beam decreased, and to assure the necessary sail-carrying power all the ballast was placed in the thick external keel – a 'lead bottom' it was originally called. Before long all the ballast was placed externally on a much deepened draught, and the boats with these lead bottoms became known appropriately as 'lead mines'. The table left tells the British side of the story.

The revenue cutter is representative of the norm of her period from which the later yachting developments emerged. The second yacht, *Mosquito*, which we have seen to have been in the fleet which met the *America*, followed the British trend of the day. By 1875, the year of *Jullanar*, the decline into extreme narrowness was steepening; but *Jullanar*'s fame is owing not to this feature but the fact that she was an early embodiment of the application of scientific naval architecture to yacht design. During the 1860s and 1870s giant strides were being made in the understanding of the physical laws governing ship propulsion, and the new knowledge was beginning to rub off on the most advanced yachts. *Jullanar* embodied the two basic principles of fast yacht design – great length and small wetted surface area. She was a lesson to the yacht architects of the day; and it came oddly enough from an amateur. *Jullanar* was designed by a manufacturer of agricultural implements. *Chittywee* represents the extreme absurdity of narrowness ultimately reached.

In 1881 the English cutter *Madge*, typifying at its best the less extreme English type of yacht, was sailing in American waters. Her influence on yachting in the U.S.A. was comparable with that of the *America* thirty years earlier in Britain.

Writing in 1885 in his *Small Yachts – Modern Practice* – a now rare volume before me at the moment – C. P. Kunhardt said 'From her advent in American waters, in the fall of 1881, may be dated a new period in the world's yachting . . . the success of *Madge* was a totally unexpected revelation on this side of the Atlantic.'

Belonging to James Coats, a well-known Scottish yachtsman who had her shipped from Glasgow to New York, *Madge* was put under the command of the professional skipper James Duggan, whose native canniness was quickly revealed. The neat little black cutter with well-dressed crew, her scrubbed white decks and coppered bottom, showed herself in a poor light indeed as she was outsailed by all the American sloops she met in casual encounters. The capital question at issue was whether the best of the narrow, lead-loaded yachts, heavy in proportion to their lengths, were intrinsically a fast type or merely fast in relation to their own English kind. That the latter type of cutter was slower than the American sloop appeared already proved before any organized races had been sailed. But Duggan was busy fooling everyone who was interested in events out in the Bay, including the newspaper reporters.

Madge first met *Schemer* in an official race, the latter one of the best centreboard sloops of her size. The dimensions of the two boats appear in the table.

	Madge	*Schemer*
Length overall	46 ft 0 in	39 ft 8 in
Length waterline	39 ft 9 in	36 ft 10 in
Beam	7 ft 9 in	14 ft 0 in
Draught (hull only)	7 ft 8 in	3 ft 0 in
Draught (with centreboard)	no centreboard	9 ft 0 in
Ballast on keel	10½ tons	4½ tons

Madge, now being handled seriously, proceeded to defeat *Schemer*, and then in further races two other reputable sloops. In a fourth race she met *Schemer* again, and in very light airs the centreboard sloop was able to show its superiority under such conditions, especially when with the wind astern the sloop was able to house her centreboard, while the cutter had still to drag the full depth of her long keel through the water. But with time allowance in her favour, *Madge* still won on corrected time. She went on to defeat another American sloop in the next match, and then came up against the sloop *Shadow*, shorter on the waterline than *Schemer* by some 2 ft but with a few inches more beam; but significantly her hull draught, excluding centreboard, was 2 ft 5 in more than *Schemer*'s. This sloop alone of the four that challenged *Madge* defeated her in a race in which the latter broke a starboard spreader and hence could not be trimmed properly on the port tack. One more American sloop remained to be confounded by *Madge*, which duly occurred; when the latter's record became an unbeaten one except for the race in which she was damaged, and all the wins but one being gained without calling upon time allowance, though she was the smallest boat involved. At the end of the contests the initially immaculate *Madge* remained acceptably immaculate, while the sloops were in various stages of disarray, the outcome of the high stresses generated by their hulls, stiff and unyielding until the moment when they might suddenly capsize, and the massive rigs required to drive them.

Unsatisfactory though the race between *Shadow* and *Madge* had been, it was rich in future augury. *Madge* was designed by the Scottish architect George Watson. He had been born the son of a Glasgow physician in the year when the *America* had won her cup. In 1873, after professional training, he had established himself in Glasgow as a naval architect making, as his letter to prospective clients said, 'the designing of racing and steam yachts a speciality. . . .' In the latter respect George Watson was establishing what was virtually a new profession, and he has justly been described as the 'father of scientific yacht architecture'. He was heir to the shipbuilders in Holland who had produced many yachts as part of their business. When in the seventeenth century Phineas Pett in England was instructed by the Lord High Admiral to 'build in all haste a miniature pleasure ship' he became the first English yacht architect; but it was

Opposite

A rare photograph of the centreboard sloop *Schemer*, beamy and shallow, 36 ft 10 in long on the waterline and 14 ft in beam, the representative of traditional American ideas in design, in races against the narrow British cutter *Madge* in 1881. 'The success of *Madge* was a totally unexpected revelation on this side of the Atlantic,' wrote an American authority shortly afterwards.

as a fractional part of his official naval duties. The profession of yacht architecture, like the craft that established modern yacht builders, was the product of the latter half of the nineteenth century, and mainly of the last quarter.

The glamour of yachting rubbed off on the yacht architect. While most of the eminent naval architects and ship designers passed their days unknown beyond their own professional circles, yacht architects might become small national figures, as did the trainers of race horses. To gain this kind of celebrity the architect needed to have created racing yachts of distinction. So it was that, in the U.S.A., Edward Burgess sprang to a fame extending far beyond the waterfronts when he produced the successful America's Cup defender *Puritan* of 1885. In Britain George Watson, designer of the royal cutter *Britannia*, became probably better known to the public than Sir William White, Director of Naval Construction to the Pax Britannica Royal Navy. Yachts were described as the poetry of naval architecture.

By 1875 there were forty-nine yacht clubs in the British Isles, but yachting possessed no overall authority. This made increasing difficulties. Without a governing body there could be no generally accepted racing rules. Leading clubs, such as the Royal Yacht Squadron, Royal Northern and Royal Thames Yacht Clubs, had produced their own sailing rules, and these were to form a nucleus for what was later produced by the Yacht Racing Association and the International Yacht Racing Union. Meanwhile many yacht owners – and they were increasing in number – who took their yachts round the coast to the racing centres would find different sailing rules on the Clyde, for example, from those in the Solent, different on the Thames and at Kingston; and in the disputes that inevitably arose as a result there was no court of appeal. The clubs had grown independently and like sovereign states ruled in their own waters. Efforts to form a governing body had been made and failed on several occasions during the decades 1850–70.

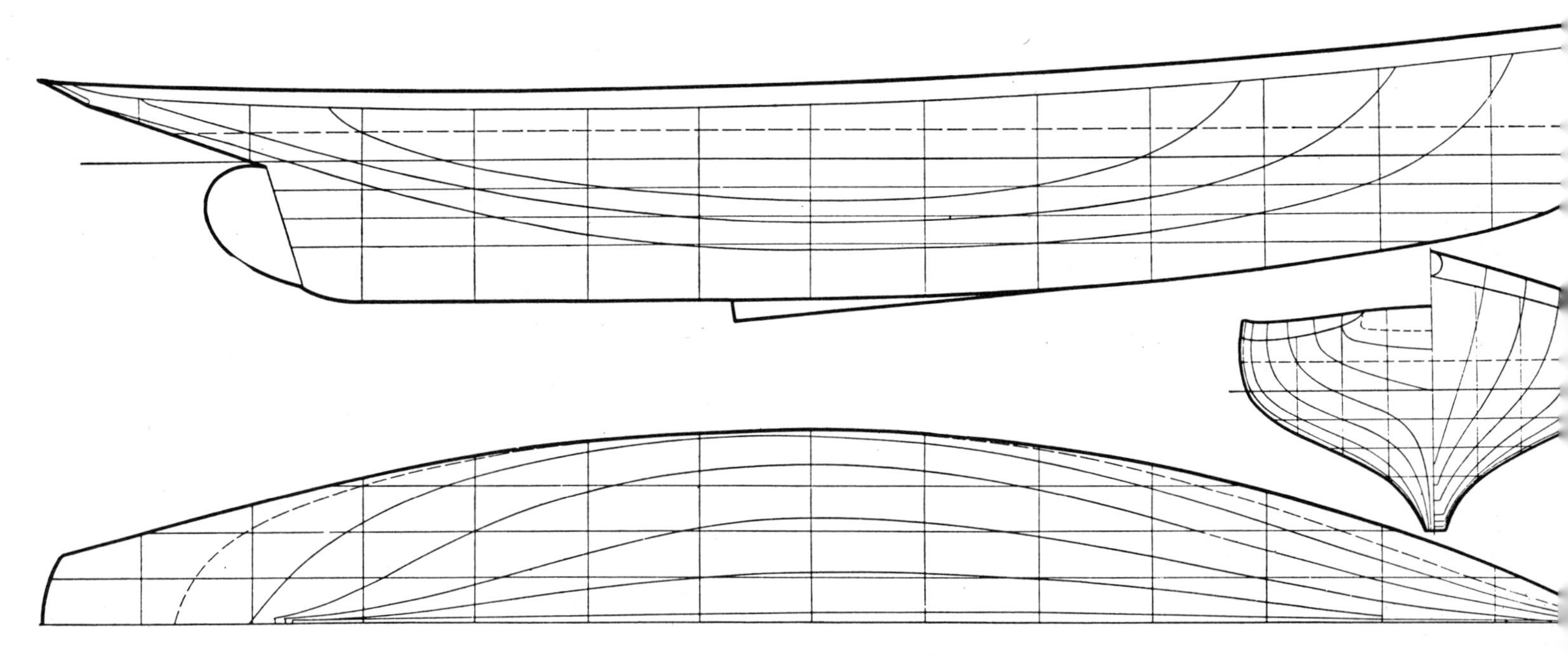

The approach to compromise in American design: the Edward Burgess designed *Puritan* was both narrower and deeper than the extreme American type of skimming dish and had an external keel of 27 tons on a displacement of 105 tons. She successfully defended the America's Cup in 1885 against the typical English cutter *Genesta*, and with her the name of Burgess became famed nation-wide.

In 1875, on 17 November, the Yacht Racing Association was formed in Willis' Rooms in St James's, London. In the chair was the Marquis of Exeter, then Commodore of the Royal Victoria Yacht Club, Ryde, Isle of Wight, supported by thirty-five representatives from various clubs. A committee was formed, of which the first President was Lord Exeter; Dixon Kemp, a leading authority on yacht racing and architecture, as well as a talented journalist and author, who will be encountered again in these pages, was appointed Secretary.

The meeting at Willis' Rooms was the outcome of a circular which had been distributed earlier in the year over the signatures of a number of prominent yachtsmen. It is stated that they were of the opinion that an association was needed of past and present owners of racing yachts above 10 tons Thames measurement, and that the association should then elect a committee, which would be re-elected annually, entrusted with the following duties:

> 'To codify existing yacht racing rules, and make such alterations and additions thereto as the committee may deem advisable.
> 'To decide disputed points connected with yacht racing. The decision of the committee to be final.
> 'To classify yachts as it may deem advisable for racing.'

The committee having been formed, it proceeded to draft a set of 'sailing rules' and a scale of time allowance for handicapping was also devised. These racing rules were derived from a study of various different sets of rules then being used by a number of leading yacht clubs, all of which in a general sense were based upon the steering and sailing rules of the Merchant Shipping Act of 1854, but suitably modified for the peculiar conditions under which yachts are raced. This involved one important change of principle. The steering and sailing rules are primarily concerned with preventing vessels from getting into

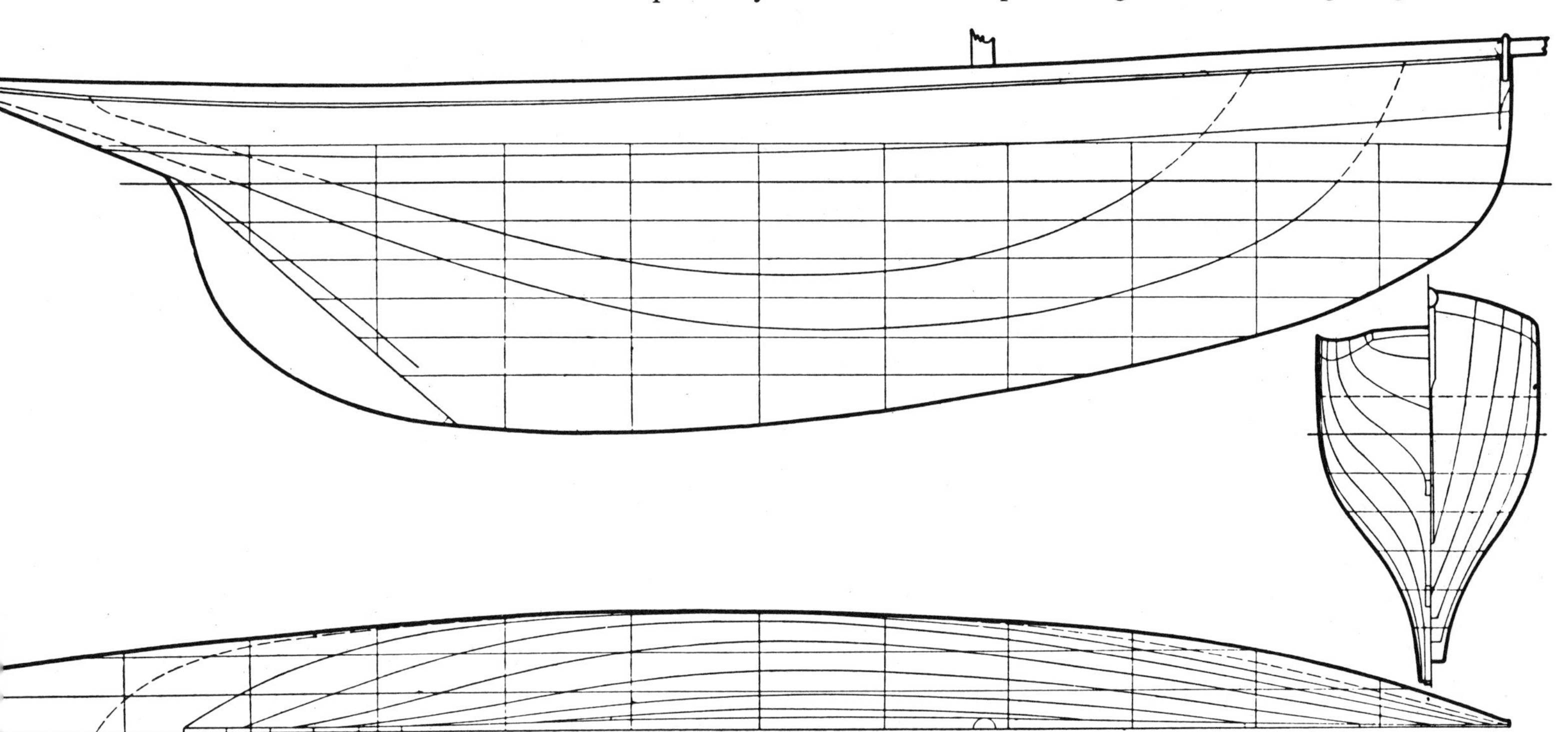

The extreme type of narrow, deep, plank-on-edge cutter produced by the British measurement rule for yachts in 1882. *Clara* was designed in 1894 by William Fife Jr, and was 64 ft 4 in in length overall, 52 ft 9 in on the waterline, beam 9 ft 2 in and draught 8 ft. More than half her displacement of 37 tons was in her ballast keel.

dangerously close quarters, and a risk of collision may be deemed to exist when two ships are far apart. Yachts race perpetually at close quarters and perform evolutions involving a deliberate risk of collision. Eighteen years after the Yacht Racing Association rules had been produced their legality came into question following the collision on the Clyde of the cutters *Satanita* and *Valkyrie II* and the sinking of the latter. The underwriters claimed the illegality of the Yacht Racing Association rules under which the two yachts were racing. The case was carried to the House of Lords and produced the famous '*Satanita* Judgement', which held that yachts in a race are legally bound by the Yacht Racing Association rules.

In February 1876 the committee's provisional sailing rules were presented to a meeting of the Association, when seventeen yacht clubs adopted them immediately, together with the time scale. But in an age happily undisciplined by overmuch centralization, when every Englishman and institution was pig-headed with proud independence, many clubs refused to join the new association; and though the latter gradually spread its authority wider, still in 1881 the Royal Yacht Squadron, the Royal Thames and the New Thames Yacht Clubs stood aside. At this point King Edward VII, then Prince of Wales, earned his popular reputation as a peacemaker, which some historians question. In 1881 the Prince was Commodore of the Royal London Yacht Club and President of the Yacht Racing Association; in the following year he also became Commodore of the Royal Yacht Squadron. With this genial figure at the head of two of the three contending organizations the moment seemed propitious for the establishment of peace between them. It was time for it; for the rancour of high yachting politics had, during the season of 1881, spread itself drably over the sailing waters. Yachtsmen who supported the Association agreed to boycott the regattas of the uncooperating clubs. This succeeded in virtually wrecking the regatta of the Thames Yacht Club, while in the Solent the regattas of the Squadron and Royal Thames presented the dismaying spectacle of Cowes Roads packed with more yachts than ever before and the racing being confined to some seven yachts dully rounding the familiar courses for the formerly hotly contended trophies. The young Association had proved itself adroit in the game of power politics. There must have been many scenes, now lost to history, full of social irony and mordant high feeling in that quaint Cowes gathering in the summer of 1881.

The dissenting clubs may have had some justification for objecting to the policies of the Yacht Racing Association. It was maintained by the Squadron that the Association unduly favoured pure racing yachts at the expense of the cruisers which formerly participated in the clubs' regattas. The Squadron produced the remarkable statistics that out of the 2,500 craft then composing the British yachting fleet, there were only three 40-tonners, two 20-tonners and a single 10-tonner that could race with any chance of winning! However, this apparently calamitous situation did not prevent the three clubs joining the Association. It occurred in 1882 not long after the Prince of Wales had arrived in Cowes one evening in August amid the thunder and smoke of guns firing the royal salute, to take up his office as Commodore of the Royal Yacht Squadron.

Two years after the foundation of the Yacht Racing Association Dixon Kemp was organizing a movement amongst leading yacht clubs for the establishment of a yacht register on the lines of *Lloyd's Register of Shipping*, the original intention being, it appears, that such a register should be an independent yachting organization. But in Britain, as in the U.S.A., a number of leading yachtsmen were connected with the shipping and shipbuilding industries, and this probably encouraged the approach to *Lloyd's Register*, which with its organization and experienced staff was uniquely well placed to establish a yacht register.

In the history *Lloyd's Register of Shipping 1760–1960* George Blake writes thus of *Lloyd's Register of Yachts*:

'It is one of the agreeable features of the history of *Lloyd's Register* during the second half of the nineteenth century that the Society, much taken up with controversy about the load line and such issues, involved in always anxious negotiation with the proud men of Liverpool, found time to establish a separate Register of yachts.

'Many readers, concerned with the fitness of the British Merchant Navy, may regard this departure as bordering on the frivolous, but it was not so. Under steady Royal patronage, yachting had been for a long time a special and elegant expression of the British genius in seafaring. The craft of yacht design and building had become an exquisite speciality, and many valuable principles of importance in the construction of larger craft was proved in the small racing boats of the inshore waters.'

There were 320 subscribers to the first *Register of Yachts*, which appeared in 1878, a number which had risen to nearly 1,000 by 1884. Included in the *Register*, then as now, was a list of British and foreign yacht clubs. There were coloured plates of their flags and burgees, a feature which was discontinued only in 1966, when this was first produced as a separate volume and not reprinted annually. In 1903 a separate *Register of American and Canadian Yachts* was published, and together with the British register it has been an annual publication, broken by the two world wars, since that date.

Initially the *Register* did not deal with classification. Shortly after its establishment scantling and constructional rules were produced for yachts of wood, iron or composite construction, and in 1888, when the first rules for steel ships were produced, they were accompanied by separate rules for steel yachts. By 1885 six hundred yachts had been submitted for classification.

The Yacht Racing Association had been founded initially with the object of bringing order amongst the often conflicting sailing rules of the various yacht clubs. Later, as will be seen, it extended its attention to the vexed question of measurement rules for racing yachts, and these rules led, early in the twentieth century, to a further step in the organization of yachting with the foundation of the International Yacht Racing Union, and the establishment of measurement rules recognized internationally.

Madge's performance was a fresh memory and the battle between the British and American types still proceeding with an intensity that shook friendships among the dedicated yachting people of New York and Boston, when attention was again turned towards the America's Cup. In their influence upon yachting in general, the contests of 1885, '6 and '7 were the most important ever to have been sailed. Thereafter cup matters became increasingly esoteric, though remaining the world's most gruelling yacht racing test. The above three contests forced the battle of the types to a conclusion in a compromise that permanently affected the design of first-class yachts; and its influence travelled to yachts of all sizes.

These contests brought to the fore Edward Burgess, a new figure on the creative side of yachting who briefly and brilliantly dominated the scene, and died young with his place secure in the story of the sport. His son Starling was also to dominate America's Cup design during the inter-war period, besides being the creator of some of the most celebrated American racing yachts of his time.

Edward Burgess's career as a yacht architect was meteoric and tragically brief. As an amateur with no formal training in any branch of naval architecture or engineering – he happened to be an adequate entomologist, in a well-to-do unprofessional kind of way – he turned to the yachting business with his brother when the family fortunes failed. In 1884 a twin challenge for the America's Cup was made, two typical English cutters of the larger class being named, the *Genesta* and the *Galatea*, the latter to challenge immediately following *Genesta* should she fail. The New York Yacht Club accepted the slight irregularity of this procedure and international relationships were a decided improvement on those

Douceur de vie afloat in 1886, when gracious living and fast sailing were not inimical. The saloon of *Galatea*, America's Cup challenger in 1886.

of the former Anglo–American engagements. This was despite American fears for the outcome; for with the lessons of *Madge* and the cutter breed so recently and firmly implanted it was evident that no yacht was available of a suitable type and size for the defence. In America, unlike Britain, the schooner remained the most popular rig for large yachts, and there was a shortage of the American type of sloop fit to meet a good English cutter. That the defence should have turned to Carey Smith for the design of a defender was not surprising. A highly trained naval architect of the new school, he was also the most experienced American yacht designer. That a second potential defender should be ordered by a Boston syndicate from the design of a gentle unassuming naturalist with no experience of large craft was bizarre.

But the much-experienced Carey Smith had his troubles. 'I've got to design a damned big scow' he told someone in disgust while he was working on the drawings for his defender. Ruffled feeling between the owners, who pay, and the architects, who create, is a theme ancient and modern in the yachting story. Burgess was able to escape this occupational hazard. He described the conception of the yacht he was designing: 'The object was not so much to get a vessel that would be particularly fast in light weather as to produce a good all round yacht, and especially one that would give a good account of herself in a breeze of wind.' This might be accepted as the credo of one who was seeking the fine compromise between the extremes of British and American design. And this he achieved.

Burgess's yacht materialized as *Puritan*, a fitting Bostonian name less reverently known in the New York circles favouring the Carey Smith potential defender as the 'Boston bean boat'. She was skippered and steered by Captain Aubrey Crocker, who had distinguished himself in the handling of the smaller sloop *Shadow*, the one boat that had taken a race from *Madge*. Against *Priscilla*, the New York Yacht Club's potential challenger by Carey Smith, and two other yachts of older vintage, she proved unmistakably the best, particularly in stronger winds.

Puritan, with less beam than the extreme type of American sloop and more draught, carried an external lead keel through which a slot was cut to take a typically American centreboard 22 ft in length and 11 ft deep. Though radically different from the English *Genesta*, she was a healthy model of vessel, and the contest was a true 'battle of the types'. Both yachts were good of their kind and the racing was close. In the light winds of the first race the centreboard had the advantage and won. In the strong winds of the second *Genesta* was in a position to win. Well ahead at the first mark, which was leeward, she sacrificed the race by hanging on to too much sail while the American sloop sent down her topsail and housed the topmast. In a strong wind *Genesta* was overpowered and possibly lost the cup as well as the race (the contest consisted of three races only at the time), for she was worthy to win.

While the challenger lost the contest her owner won immense popularity. The first race had been abandoned for the lack of wind; at the second attempt a day later the challenger was fouled at the start – a simple port and starboard tack case with the challenger on the starboard – and *Genesta* drove her bowsprit through the defender's mainsail. Her rights were indisputable, and the challenger was told by the committee that she was entitled to sail round the course alone and claim the race. The owner, Sir Richard Sutton, won the widest acclaim by refusing to do this.

It had been decided that *Galatea*'s challenge should not be made until the following year. Her owner, Lieutenant William Henn, R.N., was a retired naval officer rich with Irish charm and like Sir Richard Sutton a sportsman of the best order. Once again the social atmosphere was all sweetness and light. But *Galatea*, like *Genesta* designed by J. Beavor Webb and of a similar type, was more a cruising than a racing yacht. The owner and his wife lived on board, and the yacht was furnished heavily as a home. Webb, indeed, had left the navy at his own wish in order to devote himself to sailing, and had lived on board a previous yacht for a period of seven years. He was the type of yachtsman rare at the time which was to become well recognized in the following century – a type, however, not usual in first-class yacht racing.

For the defence Burgess had produced *Mayflower*, which after a bad start proved clearly superior to *Puritan*, and again of the compromise type. To sail her the professional skipper known to history as 'Hank' Haff was chosen, a name like that of Charlie Barr later that stands high on the small list of those who followed the esoteric profession of handling large racing cutters. In the contest *Mayflower* won two races in succession without difficulty. This contest, like the former one, aroused public interest, which had dropped so low during the Canadian challenges, and the America's Cup won the prestige which was to grow yet further during the following decades. Unhappily the good relations established were to go into sharp decline. Within nine years the America's Cup was to produce an embroglio which turned sport into an ugly international incident, and did a little to fan the Anglophobia that is ever latent in parts of the U.S.A.

5
From tradition to sophistication

In Britain at this time Dixon Kemp was the leading authority on yacht racing and yacht measurement. Yachting Editor of *The Field*, which found its way into most country houses in England; Secretary of the Yacht Racing Association; author of the two standard works *Yacht and Boat Sailing* and *Yacht Architecture*, in his later years – he died in 1899 – he gave up active sailing, and dressed like a countryman rather than a yachtsman, watched the racing critically from convenient points ashore. Other than in print he said little, though he did once observe privately that there were only two people in Britain who could write sense about yacht racing, and he was one of them.

In March 1887 he delivered a paper to the Institution of Naval Architects in London entitled *Fifty Years of Yacht Building*. In the course of it he emphasized the fact that tonnage rules had 'assisted in bringing yacht building to a standstill in this country, as no one could be found willing to build a longer, narrower and

The New York Yacht Club fleet at Newport, R.I. in 1888. The high proportion of schooners will be noticed; also two catboats.

deeper boat for any given tonnage than those which already existed, and the rule would not admit of trying experiments with beam'. The Thames measurement rule, in fact, was played out.

He was supported by later speakers, who in turn trounced the type of yacht that had been produced by the racing of recent years. Captain C. FitzGerald said that he would 'hail with delight any change in the rules which will induce yacht builders, instead of making these shapeless masses full of lead, to go in for beauty and symmetry of line, and all other things being equal I have no doubt we shall be able to sail against the Americans'. The Vice President of the Institution, B. Martell, thought that 'it is simply lamentable, looked at from the professional point of view of the naval architect, to find that the Yacht Racing Association has encouraged as much as it possibly can the worst type of vessel it is possible to conceive . . . similar to a capsized soap box'. These might seem hard words to apply to a rule that had produced such yachts as *Madge* or *Genesta*, as good in their way as the best of the American type; but in the five years since *Madge* had made her reputation in the U.S.A. English yachts had become even deeper and narrower.

However, as he spoke the spring of 1887 was about to open. Tomorrow was 1 April, and for the approaching summer's racing a new rule was to be in force. The Yacht Racing Association had appointed a committee to consider the existing measurement rule, and after obtaining the opinions of leading designers and builders, in November 1886 it had recommended the adoption of a new rule, of Dixon Kemp's devising, and destined to become famous as the length and sail area rule:

$$\text{Rating} = \frac{\text{length waterline} \times \text{sail area}}{6000}$$

This rule, in which the only taxed dimensions were waterline length and sail area, was the one that Dixon Kemp had been advocating some years before and which had been adopted by the American Seawanhaka Corinthian Yacht Club in 1883. With the drawing together of the systems of measurement on the two sides of the Atlantic, the 'battle of the types' was coming to an end with a kind of yacht compromising between the two extremes.

But not everyone was happy about the Dixon Kemp system of measurement. The Vice President of the Institution of Naval Architects immediately said so: 'I think, after the mountain labouring, as it has been, for a number of years to bring forth such a mouse as to merely take the length and sail power, and to ignore all the other principal dimensions, is a very lamentable result to arrive at.' The designer C. P. Clayton, claimed that the rule when applied to the smaller classes would produce $2\frac{1}{2}$-Raters 30 ft long on the waterline: that is, where the former rule had led to uncontrolled displacement, the new one would produce uncontrolled length. Clayton was right; and so, in his different way, was the Vice President.

No rule as simple as this, the mere product of length and sail area and without other safeguards, could do other than produce freak yachts. But still, in 1887, yacht designers had taken only the first bite from the apple of the tree of knowledge. In this game of building racing yachts to formula there was much yet to learn, and all the innate conservatism induced by the sea to restrain the wilder flights of knowledge. In the years immediately following 1887 there followed no immediate and radical change in the form of yachts, and the changes there were, especially in the bigger classes, were for the better. As usual experiments were most radical in the smaller classes. But within a decade the innocence of the

Opposite

Regatta of the Royal Northern Yacht Club on the Clyde, *c.* 1870.

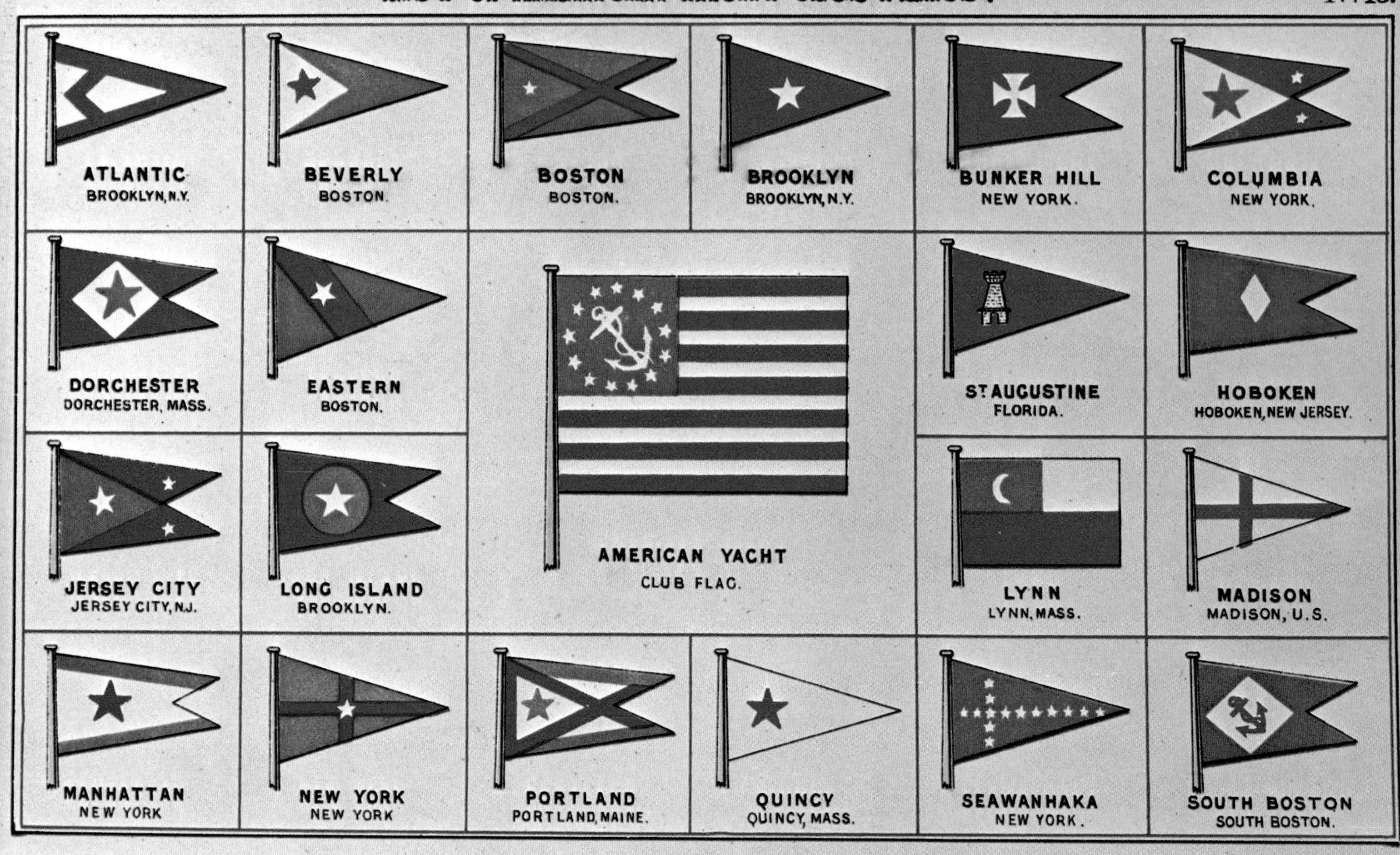

LIST OF AMERICAN YACHT CLUB FLAGS.
Nº 19.
ATLANTIC
BROOKLYN, N.Y.
BEVERLY
BOSTON.
BOSTON
BOSTON.
BROOKLYN
BROOKLYN, N.Y.
BUNKER HILL
NEW YORK.
COLUMBIA
NEW YORK.
DORCHESTER
DORCHESTER, MASS.
EASTERN
BOSTON.
AMERICAN YACHT
CLUB FLAG.
ST AUGUSTINE
FLORIDA.
HOBOKEN
HOBOKEN, NEW JERSEY.
JERSEY CITY
JERSEY CITY, N.J.
LONG ISLAND
BROOKLYN.
LYNN
LYNN, MASS.
MADISON
MADISON, U.S.
MANHATTAN
NEW YORK
NEW YORK
NEW YORK
PORTLAND
PORTLAND, MAINE.
QUINCY
QUINCY, MASS.
SEAWANHAKA
NEW YORK.
SOUTH BOSTON
SOUTH BOSTON.

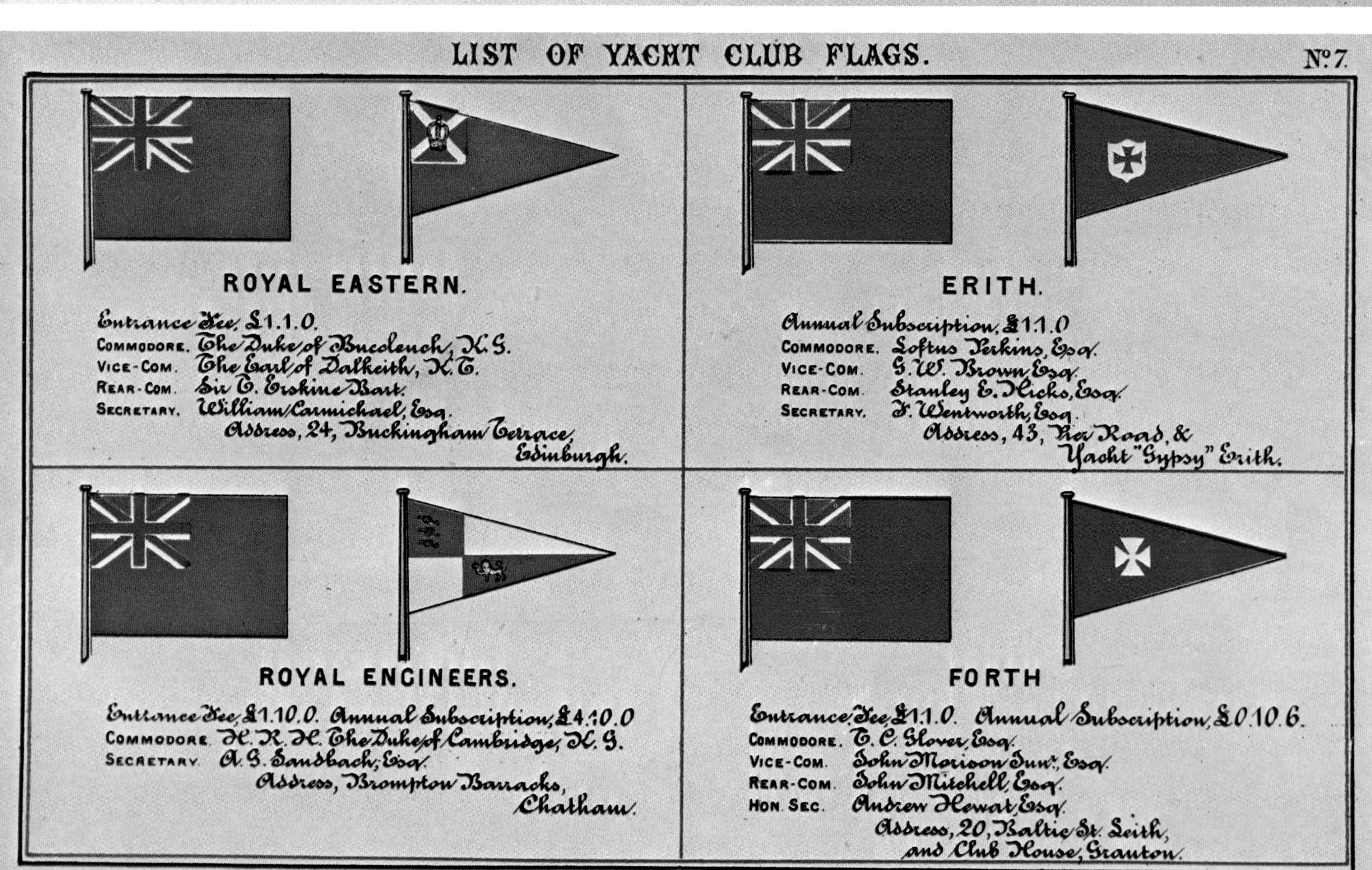

LIST OF YACHT CLUB FLAGS.
Nº 7.
ROYAL EASTERN.
Entrance Fee, £1.1.0.
COMMODORE. The Duke of Buccleuch, K.G.
VICE-COM. The Earl of Dalkeith, K.T.
REAR-COM. Sir T. Erskine Bart.
SECRETARY. William Carmichael, Esq.
Address, 24, Buckingham Terrace, Edinburgh.
ERITH.
Annual Subscription, £1.1.0
COMMODORE. Loftus Perkins, Esq.
VICE-COM. G. W. Brown, Esq.
REAR-COM. Stanley E. Hicks, Esq.
SECRETARY. F. Wentworth, Esq.
Address, 43, Pier Road, & Yacht "Gypsy" Erith.
ROYAL ENGINEERS.
Entrance Fee, £1.10.0. Annual Subscription, £4.10.0
COMMODORE. H. R. H. The Duke of Cambridge, K.G.
SECRETARY. A. G. Sandbach, Esq.
Address, Brompton Barracks, Chatham.
FORTH
Entrance Fee, £1.1.0. Annual Subscription, £0.10.6.
COMMODORE. T. C. Glover, Esq.
VICE-COM. John Morison Junr., Esq.
REAR-COM. John Mitchell, Esq.
HON. SEC. Andrew Hewat, Esq.
Address, 20, Baltic St. Leith, and Club House, Granton.

length and sail area rule had been ravished by the yacht architects, driven by the hectic competition of racing, and becoming wily in the arts of seduction. Meanwhile, too briefly, racing yachts moved through perhaps their aesthetically most beautiful phase. Passing from the extreme of the over-heavy plank-on-edge cutters of the late 1880s to the over-light skimming dishes of the late 1890s the one classical form of beauty in yacht architecture was realized and too quickly deserted by pure racing yachts, though maintained by less specialized ones.

For the third summer in succession 1887 brought a challenge for the America's Cup, and the contest had become for a while a potent influence on the future of yachting. The challenge came from the Royal Clyde Yacht Club in the name of Mr James Bell, and after some correspondence, which had reverted to a slightly contentious note, the matter was arranged. George Watson, now the leading British yacht architect, was given the task of designing the yacht, to be named *Thistle*. Burgess, almost automatically, received the commission for the defender. Now, with the Burgess compromise-type of sloop evolved in the U.S.A., and in Britain a new measurement rule that freed design from the former crippling penalty on broader beam, the rival yachts would for the first time not be of such widely dissimilar types as formerly. On both sides of the Atlantic there was occurring a renaissance in the creative side of yachting.

Thistle was a failure all the more surprising because until the first race was in progress she had seemed to be so formidable a challenger. Conceptually she was in some ways a finer creation than Burgess's *Volunteer*, which had little of the future in her design while *Thistle* had so much. As Watson wrote later in another context, 'To those able to see the beauties in a design, it matters less whether the ultimate outcome has been successful or not, and while to "the general" nothing succeeds like success, a few have a kindly sympathy and hearty admiration for those who have laboured, that *we* may enjoy the increase . . . in the mechanical arts there is often more genius displayed in a failure than a success, with this difference, that a mechanical idea seldom dies, but "blossoming in the dust" of one brain, is plucked and worn by another.' The blossom of *Thistle* was to be the '*Britannia* ideal'.

She was built under conditions of dramatic secrecy unknown hitherto even in America's Cup yachting, and when she went down into the Clyde from the slipway of D. & W. Henderson's yard on 26 April she was draped with canvas modestly veiling her underwater form. Across the Atlantic the building of the defender had only just begun, the design having been delayed until the leading dimensions of the challenger had been provided officially according to the terms of the Deed of Gift.

Thistle had a hull plated in Siemens–Martin steel, $\frac{3}{8}$ in thick and laid flush along the topsides, $\frac{5}{16}$ in thick and with the lower strakes lapped on the bottom, laid on angle bar frames supported by steel diagonal stringers. There were steel partial bulkheads and a collision bulkhead forward. The low bulwarks were panelled with mahogany inside and capped by an elm rail. Her clipper bow was rightly admired and this, the sheerline and the counter made a harmony of the profile which Watson repeated again and again in his yachts, steam and sail. *Thistle* carried channels to spread the shrouds, and her external chainplates reflected the traditional narrowness of English yachts, as the internal chainplates of American yachts reflected traditionally wide transatlantic beam. The bowsprit reefed and the halyards of the headsails were of galvanized chain.

From Lloyd's Register 1881, early examples of coloured plates showing flags and British and American club burgees.

Thistle proceeded to win eleven out of her first fifteen races, and failed to get a flag only twice, a performance which gave a confidence in the boat able to counteract the reports from the other side of the Atlantic that *Volunteer* too was sailing through a hardly broken succession of victories. *Thistle* made an easy Atlantic crossing in twenty-two days, and her hull, beamy in European terms, made a deep impression on American observers, who found it so unlike all that they associated with the breed of the English cutter. A month after arrival she was officially measured, along with *Volunteer*, in Erie Basin.

There she was found to exceed the waterline length of 85 ft that had been stated in the challenge when issued by the Royal Clyde Yacht Club, the excess being no less than 1.46 ft. This, and the resentment felt about the secrecy in which the yacht had been built, produced some feeling, and even the question whether the yacht should be allowed to race was considered by the America's Cup Committee.

Watson rather curiously described the discrepancy as an 'overlook'. The waterline length originally stated had of necessity been given before the yacht was launched, and with her low angled counter and slight overhang forward, the excess of waterline length could be accounted for by a not large additional immersion or change of trim. The error, however, was not a trivial one to occur in an important racing yacht; though Watson had said when he originally submitted the dimensions that while beam and depth were exact, the waterline length was the designed length on the drawings, adding 'When she is afloat and in racing trim I have no reason to expect that it will be more than an inch or two out, either way.' This was a reasonable statement. Furthermore, it does appear that neither Watson nor Bell appreciated the obligation, according to the Deed

Thistle, the America's Cup challenger of 1887, of not so advanced a form as *Gloriana* but a step towards the type the latter was to demonstrate so successfully a few years later.

of Gift, to specify this length at the time of the challenge. When the latter was questioned about the measurement, he said that the waterline length had been given voluntarily when submitting the required Custom House measurements, of which waterline length was not one. American opinion remained dissatisfied, a general feeling being that an architect of Watson's experience should have been able to provide a waterline length in the design stage with more accuracy than had been shown; an opinion, as it happens, apparently in line with Watson's own in his statement about 'an inch or two out', rather than this actual error of $17\frac{1}{2}$ in.

As hitherto when tense situations arose, the matter was put before George Schuyler. His decision as referee, allowed the boat to race, but was frankly critical of Watson: 'Although the variation between the stated and actual load waterline is so large as to be of great disadvantage to the defender of the cup, still, as Mr Bell could only rely upon the statement of his designer, he cannot in this particular case, be held accountable for the remarkably inaccurate information received from him, and I therefore decide that the variation is not sufficient to disqualify him from starting the *Thistle* in the race agreed upon.'

In fact 'the great disadvantage to the defender' was scarcely perceptible. Any difference in waterline length was adjusted by the system of time allowance under which the races were sailed, and in his statement before the cup committee Bell seems to have stated the essence of the contretemps: namely, that 'the exact waterline length did not seem of any importance'. It is evident that Watson had, reasonably, not regarded it as critical. He was capable of assuring the floatation

Volunteer and *Thistle* in the America's Cup match on 30 September 1887. Shortly after the start at Scotland Light *Volunteer*, astern, is working up to windward of *Thistle* and coming up on her quarter. For *Volunteer* see also p. 51, and for *Thistle* p. 74.

of a steel yacht to a close degree of accuracy, and his 'overlook' would appear to have amounted to a reasonable disregard of small alterations in trim in the course of ballasting and tuning-up a new yacht.

Light airs and some banks of fog delayed the start of the first race for two hours, and on a dull grey morning the white defender and the dark challenger made a cautious *pas de deux* amongst a dense sightseeing fleet whose loaded paddle steamers heeled, intent with observers, towards the yachts which passed and repassed, their canvas idly stirring, their way made hazardous by the crush. With the preparatory gun, it seemed to many observers, the dark yacht began slipping faster through the water than the light. She crossed the line nearly two minutes ahead. But a little cross-tacking in light airs carried *Volunteer* into the lead, and thence increasingly far ahead, until she led by a mile, and all the time pointing higher; and she added to her lead on the ten mile reach which followed. The wind increased with the distance between the defender and challenger, and the last leg under spinnaker was made in fast time. *Volunteer* was nineteen and a half minutes ahead on elapsed time at the finish. So disconcerting a result led to *Thistle*'s bottom being swept during the night to assure that she was not dragging anything.

In many respects it had not been a satisfactory race, with the crowded inside course and the need for local knowledge of currents, but there could be no question that the faster yacht had won.

The next race was postponed owing to lack of wind. When it was sailed three days later over the outside course, a leg to windward and return, there was wind, rain and a rough tumble of sea. Crossing the line ahead on the starboard tack, *Thistle* briefly had the defender out on her weather quarter, while both yachts displayed plenty of bottom as they worked out fast to windward. But while *Volunteer* was able to hold up to the wind as well as foot fast, *Thistle* slid down to leeward, and by the end of one hour and ten minutes of fast sailing, *Volunteer* was able to tack comfortably across the challenger's bow. After some cross-tacking near the Long Island shore, the outer mark was rounded, and *Thistle* was nearly fifteen minutes astern. This was a clear verdict on a fair and exacting test of weatherliness. During the run home in a wind that had risen further, both yachts crowded with sail, *Thistle* was able to pick up four minutes. But this could do nothing to efface her thrashing to windward; and as an American wrote, 'there was no skirl of pipes for *Thistle*'. And compared with the reception given to Edward Burgess and General Paine in Boston George Watson and James Bell may have found the occasion muted when the New York Yacht Club entertained them before they went home.

What had gone wrong in Watson's thoughtful conception, that its failure should have been so quick and complete under conditions which, though including only two races, were fair tests of ability under diverse conditions? There could be no question that despite his prior American studies and the attention he had been paying to the newly revealed facts of ship form and resistance, George Watson had been out-designed by Edward Burgess. Yet, as we have said, *Thistle* was scientifically the more advanced design, but the science was yet unsure of itself. *Thistle* lacked the keel surface to hold her up to windward.

During 1891–2 the technique of yacht design made more striking advances than at any time before or since. In the spring of 1891 there was nothing resembling the modern yacht afloat; by the autumn of 1892 she had arrived, and the familiar mean was established of what a sailing yacht should be like, about which design has since oscillated, at the dictates of fashion and measurement rules.

This applies to the design and construction of hulls, not rigs. Canvas was reaching outwards still rather than upwards. Long bowsprits stretched out at one end far beyond the hulls as did long booms at the other; and surmounting the squat, lower canvas were jib topsails and main topsails with yard and jackyard. But beneath sails richly old fashioned to our eyes were hulls of a kind we are familiar with; and even in rig modernity was approaching. Of the two yachts announcing the revolution, both by Herreshoff, one discarded the bowsprit, and was a stemhead sloop; though the mainsail was gaff rigged, and rigs became light and well engineered.

The new age appeared in the shapes of *Gloriana* and *Dilemma*, two yachts into which Herreshoff's originality overflowed during twelve months. He had not yet entered into the wider fame which the practice of yacht design had already brought to Burgess, who in this year was dying, leaving his so widely acclaimed place open for Nathaniel Herreshoff, whose name was to be given to the era of yachting now opening – the 'age of Herreshoff'. He had already shown, by the end of 1892, his possession of intellectual equipment differing in kind, not merely quality, from any other man who had concerned himself with the design of racing yachts.

Subsequently, as hitherto, it was the mind of the fine engineer which predominated. He designed great racing yachts in the days when it was virtually impossible to produce a vessel that at moments could not appear beautiful to the poet's eye. One may doubt, however, whether in his sailing yachts as in his earlier power craft, the appearance of his creations ever mattered more to him

The shape of things to come. *Gloriana*, designed and built by Herreshoff, shows the modern yacht form almost fully evolved – the sweep of the stem to the keel base, the full waterlines and drawn out overhang forward.

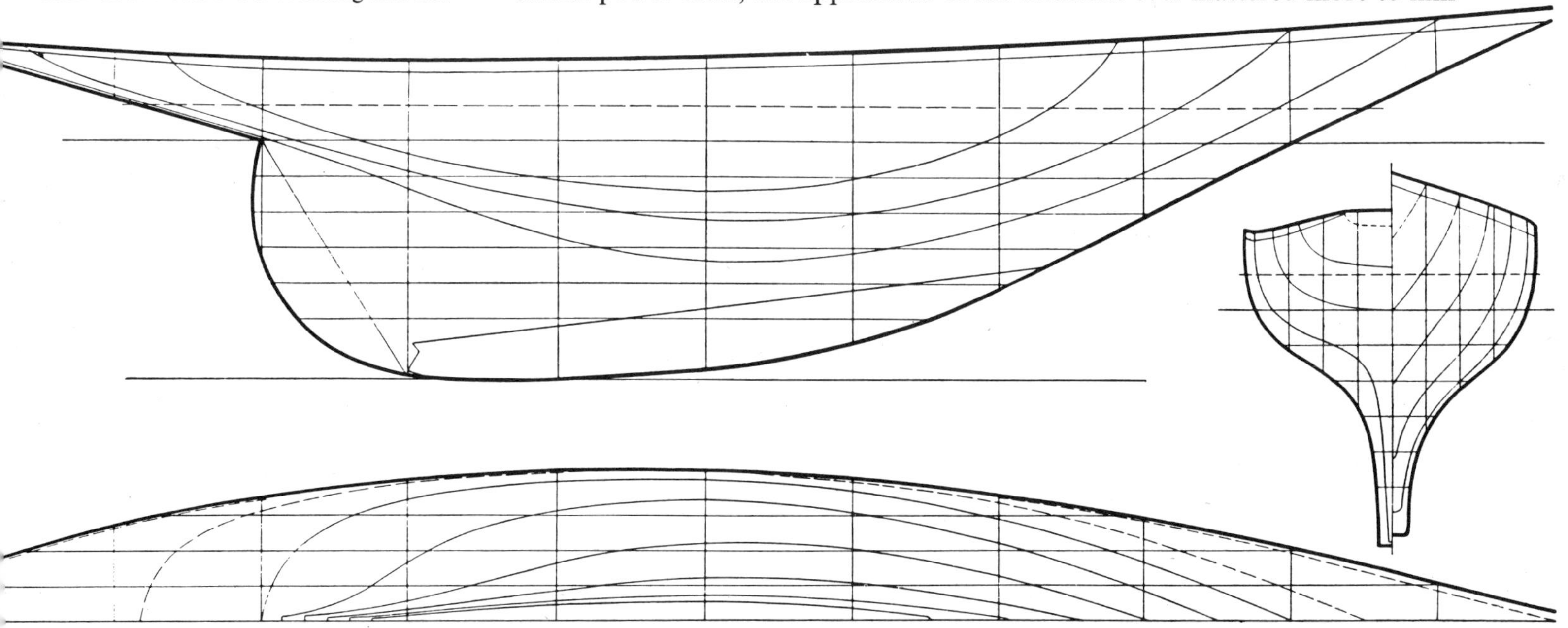

than a lavender glove to an anchorite. This makes an obstacle to the appreciation of his work today, for the engineering achievements of yesterday quickly become platitudes, while the graciousnesses of our yesterdays persist. Hence the enduring admiration for the creations of Fife, which will be loved for as long as there are people able to admire the creativeness inspired by sail; and so the coolness towards the work of Herreshoff, that most dated of all people, yesterday's brilliant technician.

With the formation of the 46-ft class to the Seawanhaka rule, Herreshoff submitted a new design to E. D. Morgan who had already been a client. This became *Gloriana*, a cutter of the following dimensions: length overall 70 ft 0 in, length waterline 45 ft 3 in, beam 13 ft 0 in, draught 10 ft 3 in. The yacht marked the re-entry of Herreshoff into sailing yacht design, and in the 46-ft class were four Burgess craft as well as one by the Scottish Fife. Before she took the water, *Gloriana* amazed and was condemned; and when she jumped straight to the head of the class she amazed and was widely misinterpreted. The early criticisms did not perturb Herreshoff, who in the autumn before that spring of 1891 had produced a smaller boat, influenced by a Fife boat named *Minerva*, but going far beyond her, and incorporating the bold features that so shocked everyone when *Gloriana*'s revolutionary character stood revealed just before her launch at the Herreshoff Manufacturing Company.

Because of her essential modernity, there is little about her to surprise today. As when the *Britannia* came out two years later for her first race on the Thames, the public eye was riveted to the bow of the new yacht, the great outreaching overhang, which was not a mere clipper stem, having only the slightest or no effect on the run-out of the forward sections, but formed an extension of the hull's body beyond the waterline. The attention focused upon this evident curiosity in the new yacht over-emphasized its importance, important though it was. Five years earlier Watson's *Thistle* had shown in American waters a bow stretched out into an embryonic overhang. In *Gloriana* herself the overhang was not fully developed, but retained in its profile a trace of concavity, as though the designer as he carved his model could not quite evade the influence of the clipper. But the overhang was far more accentuated than in Burgess's *Gossoon*, in which the clipper profile was no more than a decorative outline of stem timber; or even than that of Fife's *Minerva*. The important difference lay not merely in the extension of the forward overhang but in the shape of the sections forming it. They were rounded sections, carrying on the character of the midship body, whereas both *Thistle* and *Minerva* showed the typical hollow flare of the period.

Related to the startling overhang was the bold elimination of the knuckle in the forefoot which ran from the outreaching stemhead in almost a straight line to the base of the keel, producing a nearly triangular underwater profile. This again was more pronounced than in *Gossoon*, which had a pronounced knuckle below the waterline; but it was not more extreme than in *Thistle*.

The full originality of *Gloriana*'s form may be seen in her waterlines. 'Hollow bow lines are *not* essential to speed' was to become a well-known statement of Herreshoff's. Until *Gloriana* this idea was a heresy, and designers sought to run out as finely as possible into a knife-like entrance the waterlines as they approached the forefoot and stem. *Gloriana*'s were bluffly rounded-in forward in a way that could only suggest high resistance and a fuss of bow wave. Her smooth passage through the water was another surprise to the observers in 1891. One of them recorded at the time her small lee bow wave.

The exaggerated overhang, the suppressed forefoot, the bluff entry, achieved three crucial effects. With length measured on the waterline under the Seawanhaka rule, the shovel nose of the boat reduced waterline length by several feet, allowing an increase in sail area, while the long overhangs at both ends returned some of the length lost on the waterline once the yacht was heeled. Unlike most former yachts *Gloriana* presented an effective heeled shape, smoother and longer than that when upright. The reduced forefoot cut away wetted surface and the coarsened waterlines added to the stiffness of the hull, i.e. the stability of form.

While the public appreciated such plainly outstanding peculiarities of *Gloriana*, attributing her success mainly to the bow, and even – inspiring Herreshoff, not yet known as 'the wily Nat', to a little of his sardonic humour – to the fact that she was painted white, which was an unusual colour for yachts

Gloriana, left, the cutter which amazed when she came out in 1891. Here, in August 1892, she steps out firmly in front of the schooner *Quickstep*.

of the period, some of her most subtle excellencies were not immediately evident to the vulgar eye. Her relatively light displacement was a primary influence on her ability, achieved as it was by light construction allowing a high ballast ratio. In this fact we see another clearly modern feature of design emerging. The method of construction evolved by the Herreshoff plant from experience in building the light, fast steamers was applied for the first time to a large sailing yacht. The composite hull, with steel frames, had double skin planking laid with white lead between the inner and outer strakes, which were not caulked, on the system later to become so familiar. In this respect *Gloriana* quite outclassed the last Burgess yachts, which were both heavier and weaker.

It was the same story where the rigging was concerned. Writing on the fiftieth anniversary of *Gloriana*'s first race in the American magazine *Yachting*, H. Maxwell said: 'Without minimizing in any way Mr Herreshoff's ability, it is nonetheless true that his rival designers had shown little engineering skill in the matter of rigging design. There was no team work between their hulls and sail plans. Once the hull had been designed, all they did was to pile on as much sail as they *guessed* could be used without too many breakdowns, the factor of safety being small. Needless to say, their creations spent many hours of each summer in the shipyards. This was not true of *Gloriana*.'

Such, then, was this epoch-making yacht, at once a rule cheater, in its evasion of the measured waterline length, a fresh hydrodynamic conception in its minimization of lateral plane, wetted surface and displacement, a fine achievement in drafting with the fairness of the lines at all the attitudes the hull might take to the sea, and finally an advanced application of engineering principles to a mechanical art in which engineering hitherto had played a minor part.

Herreshoff's second revolutionary yacht of 1891 was like *Gloriana* in seeming today to be without particular distinction, so completely has the type been

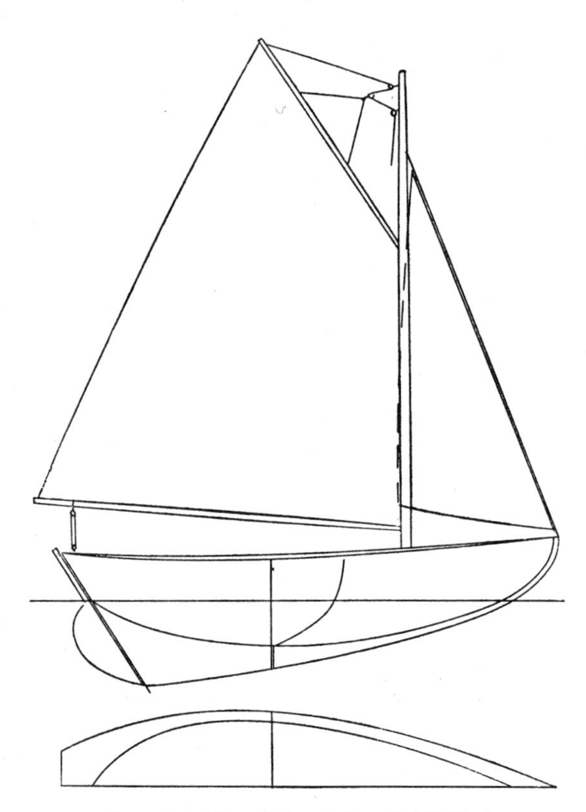

American Knockabout sloop of 1892, the mainsail without a topsail and the jib without bowsprit – the essence of simplicity with the gaff rig.

Opposite

J. M. W. Turner's impression of a regatta at Cowes (detail).

Overleaf

Royal Dart regatta in 1868. Yachts were started from their moorings or from anchor. Painting by H. Forrest.

assimilated in modern yacht architecture. The small fin-keel yacht *Dilemma* was launched in the autumn of the year and did not race until 1892, when she won every race in which she started.

The dimensions were length overall 38 ft 0 in, length waterline 26 ft 0 in, beam 7 ft 0 in, and the depth of the hull, excluding fin, a little more than a foot. Mr Herreshoff, it was said, has taken the canoe of the savage for his model. More exactly, it might be said that the hull of *Dilemma* and the successors that came fast were of the *Gloriana* form but more advanced. The hull ran out naturally into the overhangs, with fully rounded sections extending beyond the waterline and no trace of the clipper bow's concavity in the outreach of the stem. The form of the midship section was thus retained through the length of the hull in the manner that became the usual practice in yacht design; *Dilemma* and *El Chico*, her successor of the following year, could be readily accepted as new boats in 1972 rather than 1892. A century has failed to make her appear dated. This is emphasized by the comparison with the Soling One Design produced by J. H. Linge in 1966 and described as 'an ordinary boat in so far as there is nothing extraordinary about her'. This is precisely what observers did *not* think of *Dilemma* and her brood as they emerged from the Herreshoff plant during 1892.

En passant, as a commentary on rating rule influence, we might note that the Soling shows a slight reversion to the pre-*Gloriana* and *Dilemma* type in the softening of the forward waterlines and the fining of the entry. The Soling is a one-design class in which no advantage would be gained by evading measured length on the waterline. Herreshoff always worked to get the better of the prevailing measurement rule, and never more successfully than later in his America's Cup yachts.

The outstanding feature of *Dilemma* was of course the 'fin-keel', then a new term, or the 'fixed centreboard' as at first it was occasionally called; the flat steel plate of trapezoidal profile connected to the 'canoe body' (then also a new term) by two angle bars bolted through the wooden keel and carrying at its base the cigar shaped bulb of lead ballast weighing two tons. The light displacement, the considerable stiffness in the form of the hull with its full-ended waterlines, the long lever arm of the concentrated ballast, produced a powerful hull well able to stand up to its sail plan. The latter was a stemhead sloop with the then conventional gaff mainsail. When, in 1935, W. P. Stephens, the well-known authority on American yachting and an old personal friend of Herreshoff, said of *Gloriana*: 'She marked the entry in the field of sailing yacht design of Nathaniel G. Herreshoff, previously engaged mainly in marine engineering and the design of steam yachts,' Herreshoff, then in his eighty-seventh year, tartly replied: 'Since friend Stephens mentions *Gloriana* as being the product of a former marine engineer and conveys the impression that she was his original design of a racing yacht, I think it would be fair to state that I sailed my brother John's boat *Sprite* and won against the fastest and best sailed boats in the bay when twelve and a half years old, and that I began designing boats and yachts for my brother to build when I was sixteen. I also sailed my brother John's yachts in New York Bay from 1863 to 1871 and nearly always won, which is something the New Yorkers do not like to remember.'

The old man then went on to speak of *Shadow*, whose race with *Madge* has been mentioned above, 'which accumulated between 140 and 150 prizes' and he mentioned that she was unbeaten in her class until Edward Burgess produced *Papoose*. Perhaps the thought of Burgess, that rival designer born in the same year as himself who had now been dead for thirty-four years, raised uncomfortable memories. For Herreshoff allowed, as one of the rare exceptions of his life

until then, the lines of an improved *Shadow*, which he had modelled away back in the 1870s, to be published in 1935. He told of how Edward Burgess and his young wife, holidaying at the time in Bristol [Rhode Island], had become friendly with the Herreshoffs and how Burgess – a young yachtsman then of some experience, but an amateur unversed in design and construction – had delighted in this model, asking to see it whenever he came in and calling it 'the perfect model'.

And then the old retired architect, perhaps a little feline, drew attention to the similarity between the lines of the improved *Shadow* and the *Puritan*, with which we have seen, Burgess had sprung to America's Cup fame and the leadership of his profession. '*Shadow*', exclaimed Herreshoff, 'was the original of the type of model Edward Burgess so successfully carried out in *Puritan* and the following cup defenders, except that they had much of their ballast in lead castings attached to the keel, while *Shadow* had all her ballast inside.' Was Herreshoff, at that much later time regretting the day back in 1884, when Carey Smith was preparing the drawings for one potential cup defender, against *Genesta*, and the newly operating architect Burgess another, the *Puritan* herself, and an inquiry was received by the Herreshoff firm to submit an estimate for a defender. Was he remembering that the estimate had been too high to be acceptable?

Not impossibly Herreshoff's highly developed awareness of how many beans make five had delayed his entry into America's Cup racing for eight years and allowed instead the starshell of Burgess to rise. For it is by no means far-fetched to suppose that Herreshoff, with his much greater knowledge of construction and his advanced ideas mechanically, together with a sense of hull form at least no less acute than Burgess's, might have produced in 1884 the boat that would have beaten *Puritan* in the selection trials for the defence; when Burgess's incipient achievement might have been still-born.

But now, in 1891, *Gloriana* was triumphant and Burgess lay dying, and the perhaps delayed moment for the ascendancy of Herreshoff had arrived, at the crucial moment; and when it was over an age of American yachting became known to history as the 'Herreshoff era'.

America's Cup defender *Defender*, designed by Herreshoff, centre of the dispute in the contest of 1895. Painting by Antonio Jacobsen, 1897.

6 Grand yachting

With the perspective we now have of history, and with the disappearance, presumably for ever, of Big Class yacht racing, we may appreciate that 1893 and the years immediately following were the vintage time for this kind of yachting in both the U.S.A. and Britain. There was to be a Big Class again – one – when during the inter-war years leading up to the climax of the 1935 season the last of them put up a show that was a worthy final act to the lovely extravaganza of millionaires' yacht racing, the most beautiful of rich sports. But 1935 lacked the panache of 1893.

In the autumn of 1892, Lord Dunraven had ordered a new yacht to replace his first *Valkyrie* and had issued a challenge through the Royal Yacht Squadron for the America's Cup, which the New York Yacht Club accepted. A little later the Prince of Wales ordered a near sister ship to *Valkyrie II*, also to be designed by the leading British yacht architect, George Watson, who spent Christmas of that year working on the design of the new royal yacht. Early in the new year these two cutters were taking shape side by side in the yard of Messrs D. and W. Henderson on the Clyde. The ball of fashion set rolling, an order for a third yacht came from a Scottish syndicate, and she, designed by the younger William Fife whose first big yacht this was, soon lay under construction in Pointhouse, Glasgow. And while these three yachts were under construction on the Clyde, which was still the centre of the best British yacht building, far south a cutter bigger than the Clyde trio was being built at Fay's yard in Southampton. This was to be *Satanita* – the fiery and galloping 'Satan', and one of the most famous racing cutters in British yachting history.

In the U.S.A. there were five new big cutters under construction to match the British four. Here this classic season of 1893 was to be more oriented on the America's Cup than in Britain, and the new yachts were built under conditions of theatrical secrecy. This was nowhere more pronounced than at Herreshoff's yard in Bristol, where three of the new boats were being built behind locked doors and sealed windows by shipwrights sworn to secrecy. But American newspapers were a match for all the watchmen and naval standards of security, and the dimensions and characteristics of the new yachts were soon being published in the New York and New England press.

There was more variety of type among the American new yachts than the British, and more adventurous, though not necessarily successful, experiments in design. Two of the yachts were *Jubilee* and *Pilgrim*, the former designed by John B. Paine and one of the trio taking shape under guard at Herreshoff's, the latter being built at Lawley, Boston, to designs by Stewart and Binney, who were the successors of Edward Burgess. Both yachts were freakish and neither successful, though *Jubilee* did in one respect reveal the shape of things to come; though they were not to come for a long time yet. She had a fin keel and rudder hung independently on a skeg aft. This was a configuration of keel that during the 1960s was to establish itself as the normal for even the larger and seagoing

Valkyrie II sailing through the lee of the American visitor *Navahoe* in the Royal London Yacht Club's regatta of August 1893. It will be seen that the *Navahoe* has her jib topsail in stops and is about to break it out.

types of yacht. *Jubilee*'s main fin carried a bulb of forty tons of lead at a depth of 13 ft below the waterline, and through a slot in the fin and bulb a centreboard could be dropped a further 8 ft–10 ft. *Pilgrim* had a steel plate fin bolted to a canoe-like hull, with a bulb of lead fixed to the base of the plate, which was 23 ft below the waterline. This yacht has a place in history as being the largest ever to have had this form of keel, which in later years also became common. *Jubilee* and *Pilgrim* were brave experiments revealing the American technical originality which was to become such a powerful influence on yachting in later years. But it was an American – Lewis Herreshoff, brother of the designer Nathaniel – who contemporaneously said of them that 'they did not fill the mind of the true yachtsman with glowing satisfaction'. *Pilgrim* has another claim to uniqueness in yachting history. She is the only Big Class racing cutter ever to have been converted into a steam yacht!

The first boat ordered from Herreshoff's, named *Colonia*, was something like an enlarged *Wasp*, but with proportionately less draught, this having been limited by the maximum depth of water off the Herreshoff yard. Herreshoff made the same mistake as Watson with *Thistle*, only he did so more emphatically; *Colonia* was deficient in lateral area. Her poor windward work made her obviously unsuitable for selection; but some years later she was put in the hands of Carey Smith, and with the addition of a centreboard she proved fast. With *Colonia* already ordered, Herreshoff accepted, not to everyone's pleasure, a commission for another potential defender; and when it was decided that this yacht should be plated in Tobin bronze, and the small amount of this

American cutter *Vigilant*, defender of the America's Cup in 1893, and one of the nine new first-class classing cutters which came out in 1893 to make it yacht racing's finest vintage year.

alloy available was cornered for the second yacht, the owners of the embryonic *Colonia* were hardly satisfied customers.

The second Herreshoff yacht, ultimately selected defender, was *Vigilant*, and in her own way she was as radical as the others. There were the long overhangs of *Gloriana* type, but the hull was of excessive beam, a respect in which she leaned backwards to the traditional American type; while the keel, quite unlike *Gloriana*'s, gave a small hull draught in relation to the waterline length but was of great length with a toe to the forefoot. She was the biggest of that year's new cutters, English or American, with the dimensions of length overall 124 ft 0 in,

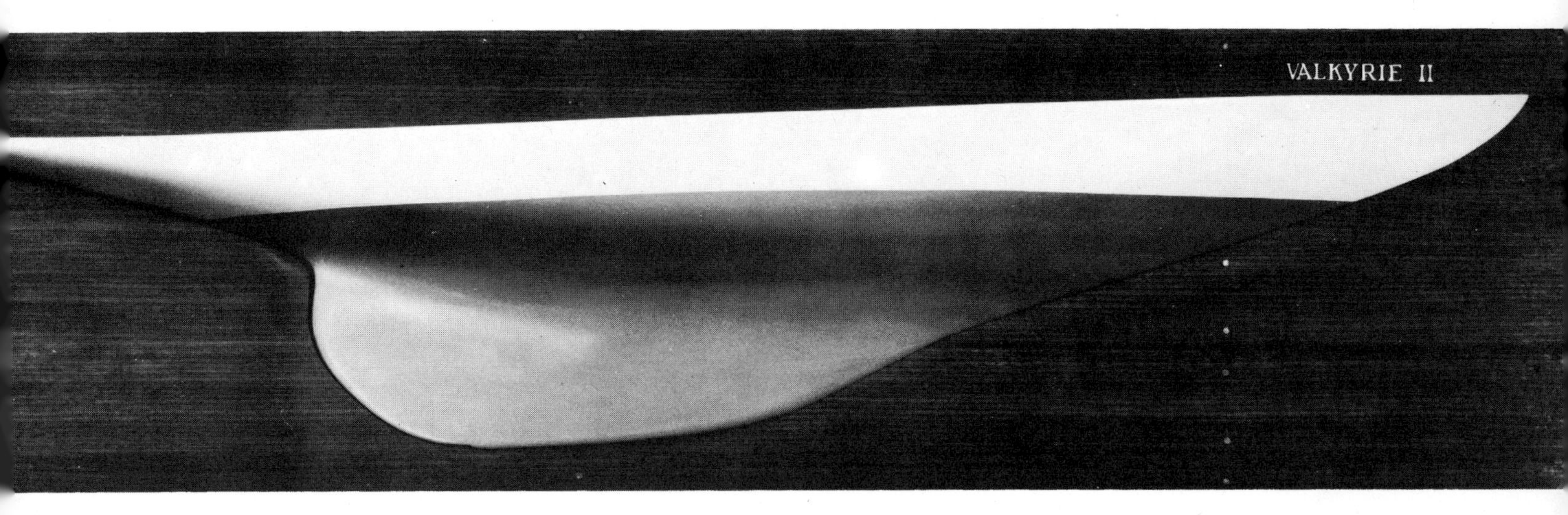

Vigilant's opponent in the America's Cup races and *Britannia*'s near sister, *Valkyrie II*. The half-model is in the Royal Thames Yacht Club, London.

length waterline 86 ft 1 in, beam 26 ft 3 in. The relatively light hull draught of 14 ft was increased to 24 ft by a bronze centreboard measuring 16 ft by 10 ft. This was of bronze plating laid on frames, cement filled and weighing some $3\frac{1}{2}$ tons. It gave trouble, not surprisingly, throughout the life of the yacht, including during the cup races. The hull, with the exception of the uppermost strakes was plated in bronze, the first time this had been used for the hull of a yacht, giving her an immediate advantage over the steel hulls of the other three with its greater smoothness and resistance to fouling. It was likewise smoother than the coppered wooden bottom of *Valkyrie*. A huge centreboard when lowered presented its considerable lateral area at the extreme fore end of the keel, this balancing a rig which had its centre of effort far forward. She was a slow boat in stays owing to her long keel with its toe, and a few years later this was rounded up in a not very successful effort to improve her handiness.

An outstanding feature of the boat was the broad beam at the deck intended to give a long lever arm to the weight of a crew no less than seventy-men strong, who became movable ballast lying up to weather. This, like her polished bronze bottom, gave her a great advantage over her rivals for the defence and also over the more normal *Valkyrie II*.

Typical of Herreshoff's work, *Vigilant* was an original and bold engineering concept able, because of her architect's technical ability, to prove successful despite the faults to which so radical a vessel must be prone. Thomas Lawson, who was later to own an even more extreme large cutter by Herreshoff, the *Independence*, summed up *Vigilant* not unfairly in the words: '. . . she was the prototype of a vicious kind of yacht, whose existence has been more of a curse than a blessing to the sport of yacht-racing.'

This should not be regarded as a criticism of Herreshoff as a designer. He built yachts for a specific purpose – such as *Vigilant* to defend the America's Cup – under the rules then established which permitted such yachts to appear.

Herreshoff was always a little more clever than the rule makers. And in the season of 1893 the two other architects concerned were producing yachts no less extreme and radical, though as experiments less successful. This year shows most vividly the contrasting national mentalities of the British and Americans beautifully expressed in eight great, extravagant racing cutters. It might be said that on one side of the Atlantic the architects were creating poems, on the other establishing theorems. There can be no question that *Vigilant*, *Pilgrim*, *Jubilee* and *Colonia* were mechanically more advanced conceptions than *Britannia*, *Valkyrie II*, *Satanita* and *Calluna*, reaching farther towards the future than the gracious British yachts, which were more reliable, seaworthy, balanced and, under some conditions, faster ships. Herreshoff himself, as we have recorded, believed that *Valkyrie II*, despite winning no races in the cup contest, was faster than *Vigilant*.

Navahoe, unlike the other four American cutters, was not designed as a potential America's Cup defender, and indeed there were elements of the cruising yacht in her design and construction which detracted a little – though this was not often appreciated – from the victory of the British yachts over her when, later in the year, she visited Britain. Her owner had announced his intention of bringing his new yacht to British waters, chiefly with the object of regaining the American Brenton Reef and Gordon Bennett Cups; so while *Valkyrie II* was to be pitted against the best American Big Class yacht in American waters, *Navahoe* was to meet the best British yacht, *Britannia*, near sister to *Valkyrie II*, in British waters. And in each case the visitor was worsted.

Never had yacht racing drawn such public interest on both sides of the Atlantic as when the fine blue spring of 1893 brought the new yachts out to their first starting lines. Summer weather in Britain that year was perfect and long remembered. Three of the new British cutters – *Satanita* was not yet in commission – sailed their first race in the Thames on 25 May under the burgee of the Royal Thames Yacht Club, supported by the older *Iverna*, clipper bowed, deep sectioned and intermediate in type between the tonnage rule plank-on-edge craft and the new type just to make its début in Big Class racing. This Thames race, over a course that has long been abandoned for yachts, should be considered in a little detail, for it ushered in the new era in sailing performance and nothing was quite the same after it.

There was a haze hanging over the Thames as the yachts came up to the line in the Lower Hope, where crowds were thick on the shore, while bands played and tugs blew their sirens. The lessons this race was to teach, and in so doing to reveal the path of the future, was thanks to the presence at the regatta not only of *Iverna*, a first-class yacht of a slightly older type, but also of the nine-year-old *Irex*, renamed *Mabel*, racing in a different class, which was representative of a yet older type, the straight stem, narrow yachts produced by the defunct tonnage rules.

As the Big Class crossed the line and set off on a two hour's reach down-river to the Mouse Lightship, those on shore and in spectator boats watched the new yachts with mixed feelings. As I wrote elsewhere of this occasion:

'On 25 May 1893 certain yachtsmen were shocked to the depths of their salt-caked souls by an appalling sight – by *two* appalling sights – in the Thames below the Lower Hope. Some of them were able to gasp through their agony, "hideous machines", or those with a taste for more syllables, "gratuitously ugly", before begging weakly for more stimulants . . . They kept saying "hideous" to themselves and one another, but this was simply because they were inarticulate

Englishmen, true to the tradition of the stiff upper lip, and not adept in the language of the feelings. Only great tragic music – fortissimo and furious – could take to mate the turmoil in their breasts.

'It will be appreciated that nothing is exaggerated when it is learned that these hardy, Victorian nautical gentlemen had just looked for the first time upon the bows of *Britannia* and *Valkyrie II*.

'It added to the offence that nobody knew what to call the desecrations. Viking bow, spoon bow – these were possible names. But obviously the language of the sea could not be expected to cope with bows that were like proper bows turned inside out. And thinking thus the old gentlemen's eyes would rest with relief on the familiar sweet curve of *Iverna*'s clipper stem, and further note with relief that she was not getting indecently far behind the new yachts in this fast reach down the river. They probably missed altogether the still, small voice of Dixon Kemp saying ". . . the eye of the rising generation, untrammelled by comparisons, will grow to love the Viking stem, just as the past generation did the swan stem"!'

Of no less significance to many watching the race than the relatively good performance of *Iverna* was the fact that when rounding the Mouse Lightship the old-style *Ilex* (fast yacht on a reach) was only about forty-six seconds behind *Britannia*, which was leading, after twenty-five miles of racing. Had *Ilex* been competing in the same class as *Britannia*, she would at this point been well ahead of the latter under time allowance.

It was when the yachts rounded the mark and came on to the wind that the

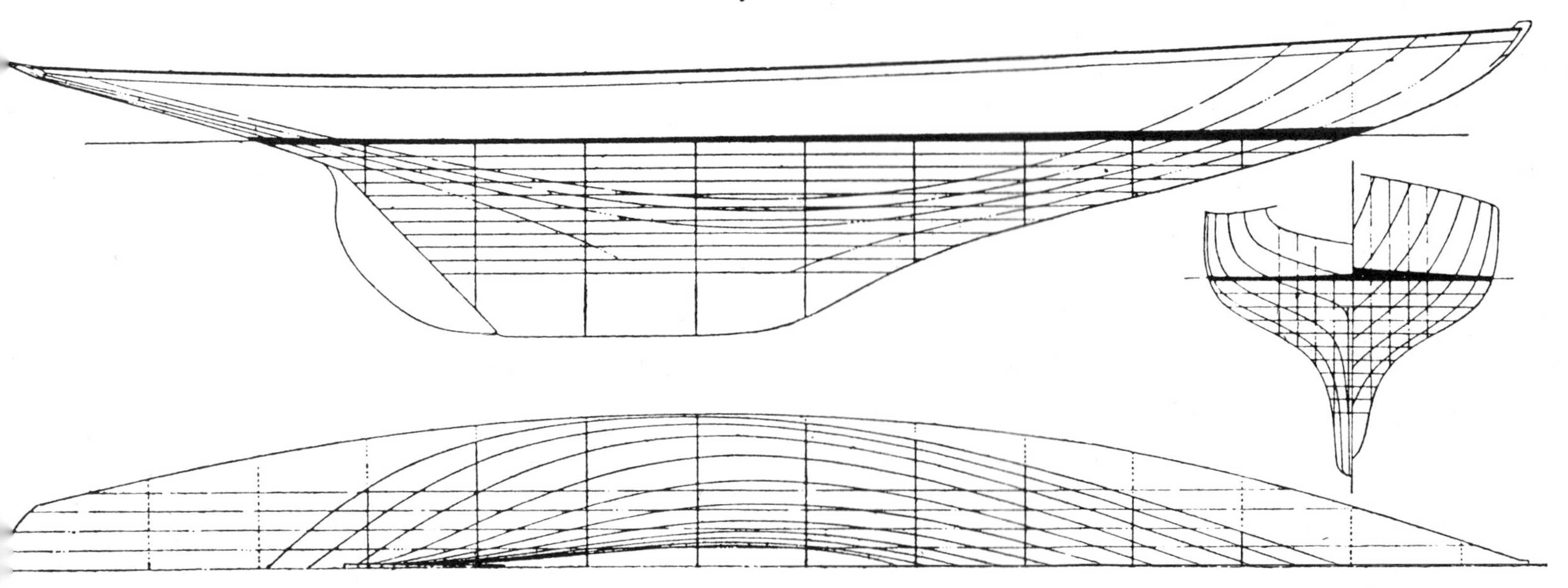

The hull form that was being approached by Watson in the *Thistle* of 1887, and was almost attained in the *Gloriana* of 1891, reaches its full expression in the *Britannia* of 1893, also by Watson, and known for generations to come as the '*Britannia* ideal'. Forty years later the basic form and proportions of yachts large and small echoed those of *Britannia*.

revelation was vouchsafed. Then it was that the new cutters began steadily drawing away from the old ones. The yachts beat back to Gravesend with *Iverna* steadily dropping farther astern and to leeward, while *Britannia* and *Valkyrie*, the two yachts remaining close together but the former keeping the lead, gave a demonstration of weatherliness that was new to the world of sailing. Between these two the race was still in doubt when, a few miles from the finishing line, *Valkyrie*'s bowsprit snapped in a squall, leaving the race to *Britannia*. From the moment that the boats had come on to the wind at the Mouse it had been clear that one or other of these two yachts could hardly avoid winning the race. The new era had opened.

Then the Big Class swung into the summer rhythm of 'round the coast' racing. In 1893 it was as it had been before and as it would continue to be, though not invariably every season – there were the years when the Big Class was in a state of collapse – until the summer of 1935, when the last strong Big Class went 'round the coast'.

This was a uniquely English feature of yacht racing, bringing the few biggest and most famous racing cutters of the day to the waters of the crowded coastal resorts. The Down Swin races, in the sandy shallows off Southend and Harwich, followed those Up Thames; thence to Dover and then the first visit of the season to the Solent. Afterwards came the Clyde races, in the loveliest of all waters, many believed, graced by the Big Class; and next those of the Irish coast, in Dublin Bay and in Cork Harbour, when the fleet sailed off Haulbowline Island, where the first yacht club had been founded, out to the Daunt Rock lightship. Returning to the Solent, the class found the yachts gathering for Cowes Week, with the naval Guardship, a sprinkling of royal yachts, the Roads busy with steam launches, the river thick with the smaller craft. But the climax of the Week did not end the season. The yachts went on to the West Country, to Torquay, Plymouth and Dartmouth. In some years they would go over to the French coast, where the racing was considered to be not of the best, but the atmosphere delightful.

Those summer months were something of an endurance test for yachts and men. The passage-making between ports off coasts that so often belied the name of summer was hard on yachts' hulls and gear. The paid hands, in the intervals between days of hard racing, were handling their craft round the coast while making good deficiencies in sails and equipment. Then they went round the familiar courses again:

> Racing every day of the week,
> Every day but Sunday.
> We finishes up of a Saturday night
> And starts again on Monday.

The arrival of the Big Class at the various ports brought glory to each local regatta, added a flush to the holiday-making spirit of the towns' summer visitors, brought prestige to the Mayor and Corporation and profit to the local tradesmen. The Big Class were the star performers amongst a big cast of local yachts, with the responsibility of providing the crowning spectacle for the civic bonanza on shore. The courses sailed by the yachts, particularly the arrangement of the starting and finishing lines, was determined partly for the benefit of spectators ashore. Inevitably this did not always improve the racing conditions, but it did bring the best known racing yachts into touch for a moment with the wider public, which did not forget them. It was many years later, at the time when this kind of yacht racing was about to go under the harrow of World War II that the American owner of the last big American cutter to race in English waters nicely described the effects of starting and finishing such races:

'The race at Dover on June 10 initiated us in the odd method of starting English races, which is so different from the system in this country. Our starting instructions read that when a certain lamp post on the end of a certain breakwater came in line with the east edge of the tower of St Mary's Church that we would be on the range, adding as an afterthought that there would be a stake boat a quarter of a mile offshore, but that this had nothing to do with the line. Reading these instructions impressed me with the fact that to succeed in racing in England one must have a knowledge of lamp posts (possibly from nocturnal prowlings) and a thorough familiarity with ecclesiastical architecture. In spite of this intellectual confusion, however, I see by a clipping from the British press, that we got the best of the start.'

Thus wrote Gerard B. Lambert, owner of *Yankee*, of what he described as 'an adventure in the summer of 1935'. We have already reached the day of the grandchild for whom he wrote his tale, and that child, an adult now, may never have seen such yachts racing as *Yankee* and her like.

But we must return to the high summer of 1893 from the Indian summer of 1935, when great yachts yet amazing and outlandish to those who had not yet had time to become accustomed to them, were out racing for the first time.

Their more extreme features were due to Nathaniel Herreshoff more than anyone else. We have seen in the life of Edward Burgess the amount of public

One of the best known of professional racing skippers, Captain Charles Barr whose talents became an integral part in the story of the America's Cup defence.

interest amounting to ballyhoo centring in the U.S.A. on the designers of America's Cup defenders, outstripping in New World intensity anything known in Britain. The unique qualities of Herreshoff's genius were revealed, his talents well proved, before he touched an America's Cup defender; but the fame that will hang to him for as long as anyone is interested in the creation of fast sailing craft, glows from the period 1892–1920, the twenty-eight years during which the America's Cup was defended six times by yachts to his design, while no other architect produced a defender. In that time, against challengers by three British architects – G. L. Watson thrice, William Fife twice, Charles E. Nicholson once – there were twenty individual races in the six contests and eighteen of them were won by Herreshoff boats. Until 1920 not a single race was lost by a defender.

The bare record of the Herreshoff era of Cup defence is thus:

1893	*Vigilant* against *Valkyrie II* by Watson.	Wins 3 races to 0.
1895	*Defender* against *Valkyrie III* by Watson.	Wins 3 races to 0.
1899	*Columbia* against *Shamrock II* by Fife.	Wins 3 races to 0.
1901	*Columbia* against *Shamrock II* by Watson.	Wins 3 races to 0.
1903	*Reliance* against *Shamrock III* by Fife.	Wins 3 races to 0.
1920	*Resolute* against *Shamrock IV* by Nicholson.	Wins 3 races to 2.

In the U.S.A. the competition during those months as the 1893 season opened was frenetic and secrecy mad. Lewis Herreshoff has thus described it: 'All attempts by the designers to keep their work secret were utterly futile, for all essential information as to dimensions and chief characteristics found their way into the newspapers, giving to the public interest and discomforture to the builders, who set seals on the mouths of their workmen and watchmen, and blocked every door and window where the prying public might steal a view of the coming wonder. . . .'

By the time *Valkyrie II* withdrew from English racing and prepared for her Atlantic crossing and the America's Cup contest, she had won eleven first prizes to *Britannia*'s eight; but in terms of winning flags – first, second and third – *Britannia* had sixteen to *Valkyrie*'s fifteen. *Valkyrie* could reasonably claim to be the best cutter in English waters by a small margin, while the two cousin ships were far out ahead of all others.

In the U.S.A. *Vigilant* was the almost inevitable selection for the defence. The contest showed that, under America's Cup conditions, the best American cutter was slightly faster than the best British. She took three straight races. In the third of them, sailed in strong winds, *Valkyrie II* had what seemed to be a winning lead at the windward mark, a tribute to her superb weatherliness. But *Vigilant* overtook the challenger on the run home by an exhibition of fine seamanship when the reefs in the mainsail were shaken out and the thimble-headed topsail was replaced by a small jackyarder. Though the challenger had not won a race, it had been a worthy series between two of the best cutters in the world. Lord Dunraven was determined to challenge again, which he was to do with unhappy results in 1895.

Back in England *Britannia* was racing a series with the American *Navahoe*, which was closely followed by the public. One more race in that year demands mention; that for the Brenton Reef Cup, to regain which was one of *Navahoe*'s

Opposite

The Big Class in English waters, 1893: left to right, *Navahoe*, *Calluna*, *Iverna*, *Santanita*.

objects; it had been won by *Genesta* in American waters in 1885. The course was 120 miles from the Needles round Cherbourg breakwater and return, and so over an offshore, not a regatta, course. The wind was strong, and the two yachts, *Britannia* and *Navahoe*, started with reefed mainsails and thimble-headed topsails, not jackyarders, set above them; and they were soon sending down those topsails. But neither yacht, in the seaway, could house their topmasts, an unsafe evolution until they were within the shelter of the Cherbourg breakwater. By this time *Navahoe* had a lead that was a matter of seconds. There was no time allowance. The yachts set back on the homeward leg with topmasts housed, and later staysails were lowered, in the wet and thrashing sea conditions that later yachts of their size were unable to face. Technically *Britannia* was the winner by a tiny margin; but in such a race as this there could be no winner. It was a matter of two magnificent yachts finely sailed in a contest of brawn, muscle and 'know how' of tough seamen, in a rivalry that raised itself to a poetic level. Grecian epics may have been made from lesser material.

That race is historically memorable and deeply significant in the story of yachting development. The great cutters of the 1893 season, the most perfect sailing machines yet devised with a weatherliness hitherto unknown, had not shed the older quality of the good ship – rugged seaworthiness. Technical developments of the future were to produce yachts similar in size, even closer winded and very much faster round the average regatta courses in light to moderate winds. *Britannia* herself forty years hence was to be out racing with this later breed. It was one from which selective breeding had eliminated the fundamental quality of ruggedness in order to acquire more specialized abilities. The great cutters of the inter-war period which became *Britannia*'s later young exquisite rivals, with subtly engineered rigs, could not have faced that Brenton Reef race of 1893, and were compelled to stay at their moorings, even when the racing courses were in wholly sheltered waters and only thirty miles or so in length, in wind strengths much below those taken in their determined stride by *Navahoe* and *Britannia* that day. When yachts again raced with confidence over long courses, taking the weather that came, it was the offshore racing yachts, comparatively tiny in size, of a breed not fully developed until after World War II.

It must be regretted that by then social and economic circumstances had thrust aside Big Class yacht racing. Yachts with the technical qualities of an ocean racer of the 1970s and the size of *Britannia* would have been glorious sailing creations. For mere size in a ship, as in a building, may be a source of inspiration not merely of wonder for the vulgar.

In the following season, 1894, the successful America's Cup defender of the previous year paid a visit to British waters to offer a sterner challenge than *Navahoe*. *Britannia* defeated *Vigilant* in the first seven races of the season and lifted yet further her reputation as a peerless cutter. In all *Britannia* met *Vigilant* seventeen times and won twelve of the races. It is of interest, in the light of both former and later Anglo–American yachting, that while the *Vigilant*'s light sails were superior to *Britannia*'s, the latter's working canvas was clearly the better. And it is of no less interest, in considering the sporting ethic, that the well-known Cowes sailmaking firm of Ratsey and Lapthorne were not prepared to make a new suit of lower canvas for the American.

None of the meetings between *Vigilant* and *Britannia*, which were followed in the newspapers by people far inland, had the dramatic appeal of the second, the race for Queen Victoria's Cup on the Clyde. A recollection of this race, printed in an article by James Meikle in the London *Yachting Monthly* of August 1924, should be preserved:

'The best race I ever saw *Britannia* sail; the best race I ever had the pleasure in a long journalistic career of seeing any boat sail; moreover, the sort of race which a novelist of genius would have been acclaimed for inventing, was that for Queen Victoria's Cup at Hunter's Quay, under the auspices of the Royal Clyde Yacht Club on Saturday, 1 July 1894.

'The weather was simply ideal in every way for such a purpose, and the matchless scenery by which the course is so charmingly framed was looking its loveliest.

'Although *Britannia* led for the first seven-and-forty miles of the fifty forming the course, the sailing had not a dull minute about it, but it was when these last three came to be encompassed that the real thrills began to operate, and the colossal crowd went half-crazy with enthusiasm and excitement. The last markboat to be rounded was that moored off Kilcreggan. *Britannia* had it first, but by so little that she had to give way to the extent of making room in which *Vigilant* could be gybed. *Britannia* would lose about three-quarters of a minute in and through the severe blanketing from *Vigilant*'s great mainsail as it was swung over, and the chief point now resolved itself into the question: Could she run through water fast enough to save as much of her allowance (three minutes) for difference of rating? Well, she did, moreover, with one minute seven seconds to spare, and perhaps in all her long and brilliant career she never sailed better than she did over those crucial three miles, and it was meet that she should, as it was estimated that the saucy sloop from over the Western Ocean covered the distance at the rate of about twelve and a half miles per hour. The sailing was of the clean full-and-by-order, and as the grand cutters went ripping at the pace of railway steamers through the smooth and sun-kissed waters with their great sails standing as beautifully as if they had been carved out of ivory the sight was one in every way worthy of the two great maritime nations represented.

'Lest it should be thought that too much rein has been given in these notes to open fervid nationalism, or worse, a whose-like-us sort of Clyde clannishness, let me put it in a few lines from the introduction of the admirable account of the race which that great English yachting journalist, Harry Horne, wrote for *The Field*:

'"Yacht-racing caught on in the South of England last season more than ever before," says Horne, "but it is doubtful if a match of any sort would ever attract such a multitude of people as lined the shores on Saturday last from Kirn to Dunoon and filled the steamboats to view the contest between *Vigilant* and *Britannia* for Her Majesty's Cup. It is true that the racing was stirring enough to have galvanized a mud engine into life, but such unbounded, delirious enthusiasm as was displayed when *Britannia* won we should never have expected from sober Scots folk. There were many reasons, however, for this. *Britannia* is a Scottish boat owned by the Prince of Wales, and the Scots are nothing if not loyal; further, she had defeated the representative yacht of America, which, as a racing vessel, was claimed to be the Queen of the Seas; but for the moment it was perhaps the keen battle which had been fought by the pair which was playing on the feelings, and was uppermost in the mind, and the outlet was the scene indescribable which followed *Britannia*'s victory. In a long experience we have never witnessed a finer race, nor one sailed with better judgment, and a meed of praise is alike due to victor and vanquished."'

It is a worthy yachting scene that can evoke such memories thirty years after the event.

In 1895 Lord Dunraven returned to the America's Cup and made what was hardly less than an indecent assault upon it. Sport has been described as 'war without weapons'. Ratsey, refusing to make sails for *Vigilant* in 1894, was a gentle expression of this attitude. Several of the sports in the Olympic Games of the twentieth century's latter half have made the crudity of the attitude plain – also perhaps its inevitability. There are those who oppose international sport because it so easily degenerates into war without weapons. The Royal Ocean Racing Club was adopting this attitude in the 1930s when it set itself against Olympic ocean racing. Unhappily yacht racing had, by this time, become the

On board a first-class racing cutter: the yacht is running before the wind, and a proportion of the crew is concentrated aft.

first sport to raise ugly international relations. It was in 1895. Yachting at this moment becomes a social study in the broad sense. The sporting ethic is ideally one of the perfect rulers of human behaviour. It is not a simple one, but contains several contradictions which meet at points around which are areas of subtle doubt.

The true sportsman must be determined to win by any possible fair method. But the meaning of 'fair' in this sense cannot be defined. It is not the mere crudity of sticking to the rules, of not cheating. This is the first rudiment of sport; and sportsmanship begins where rules end. It entails at its best a complete lack of personal vanity, a discarding of egotism. The true sportsman has to be imbued with the will to win, yet feel assured in his mind that victory should go only to the best player – where yachting is concerned, this means the best crew/yacht combination. There should be no joy in the fluked victory. Oddly, there should be no resentment about a fluked defeat either. Victory or defeat in themselves should mean nothing; only the course that leads up to either. Better a brilliant, closely contested lost race than a too easy win. Better to win after a struggle than a walk-over, however much the latter may emphasize a creditable superiority of performance. It may be all right for lifemanship, or even gamesmanship; it is not good enough for sportsmanship.

Inevitably, the more important a sporting event is, and the more it is publicized, and the greater the sacrifice in energy and money that has been entailed by each contestant, the greater becomes the stresses that sportsmanship has to withstand. It is because sportsmanship is liable to be crushed under them that many people object in principle to international sporting events.

It is yacht racing that has produced the most resounding crash of sportsmanship ever yet experienced in any sport, an occasion when a little matter of a few yacht matches became an unpleasant international episode between two of the most civilized peoples on earth.

There was a light touch of autumn in the air off Sandy Hook that September morning, where a gentle offshore breeze fanned a light haze seawards. Was there something else in the air, too? A premonition of trouble? For a few among the thousands of observers in the congested spectator fleet the trouble soon gathered itself on to the palpable shape of the large white yacht, the British Milord's yacht, which was approaching the scene under mainsail and jib only, and with a strange lack of activity on her deck, considering she was about to start a crucial race.

She had already lost two races in this America's Cup series. This would be the third and last if she lost again, for only five races were held in 1895. Though the preparatory gun had been fired the state of calm on *Valkyrie III*'s deck was unruffled; the wind was light, but her enormous jackyard topsail, which had been attracting so much attention during the last days, remained in stops; and even her staysail was unhoisted. Shortly afterwards, and still so rigged, she crossed the starting-line, two minutes after the starting-gun, and about a minute and a half astern of her rival *Defender*. Within another minute her huge tiller had been jammed down, her racing flag had been lowered, and she was heading towards her tug and a brisk tow back to port.

One might imagine a giant indrawn breath of pleasurable sensation swelling through the excursion fleet, whose bloated proportions were due to an ache for drama rather more than an intelligent appreciation of a sailing match. So the Britisher had quit, as the newspaper rumours had suggested he might! Well, New York would not miss him now – New York, which had been finding him the rage for some weeks, aping his clothes, his manners, even his odd shoes, one hard and one soft, which an attack of gout had forced him to wear briefly. But now he had quit. More than half a century later an American historian of this event could still feel bitterly enough about it to write of Lord Dunraven 'who came equipped with a chip on his shoulder, a gift for dubious sportsmanship, and a big bag of sea-lawyer's tricks with which he obviously proposed to match any that the defenders might pull on him.'

Lord Dunraven.

The first race in this unfortunate America's Cup series had resulted in a win for the defender, followed instantly by a letter from Lord Dunraven to the Race Committee making two complaints. The first was not unjustified. It concerned the obstruction of the starting-line by the spectator fleet, and the confused water through which the challenger had to sail throughout the course – for the yacht from England was the cause of greater curiosity than the home vessel, and hence was followed more closely. This crowding of the course was a chronic condition then in America's Cup racing, for there was no official patrol of the course and the sole custodians of order were yachts of the New York Yacht Club, volunteers for the task and with no real authority. Dunraven knew this, but sympathy was due to him in his frustration; and some, too, for the Race Committee, for no one had yet become accustomed to yachting as a great spectator sport. Thanks to Dunraven's forcible complaint, by the time the next challenger – Lipton's first *Shamrock* – raced, the course was patrolled by the U.S. Navy. But it might have been due to Dunraven that the chequered history of the America's Cup should have ended for ever with him.

For his second complaint that evening was nothing less than an imputation of fraud; namely, that during the night before the race, and after the two yachts had been officially measured, additional ballast had been put aboard *Defender*, which had sailed the race on a waterline three or four feet (he suggested) longer than her official measurement. Lord Dunraven was, however, still courteous in not suggesting that the reballasting had been done with the knowledge of Mr Iselin (who was in charge of *Defender*) and of the America's Cup Committee. But Lord Dunraven requested remeasurement and threatened to discontinue racing otherwise.

The two yachts were remeasured next day and the results agreed with those originally obtained.

So the second race was sailed. This began with a collision between the two yachts in which *Valkyrie* damaged *Defender*, proceeded to a sail round the course and crossed the finishing-line first, and ended with a protest case in which *Valkyrie* was disqualified. It was altogether a most ungracious affair.

The collision occurred just before the start and did not involve any complex tactical problem. There was photographic evidence showing the positions of the two yachts at intervals before and after the collision occurred; it all happened in full view of the Committee boat; and the facts have never been in dispute. From the point of view of the then racing rules *Valkyrie* was clearly in the wrong – a view generally accepted by everyone since.

Valkyrie's boom-end ironwork fouled *Defender*'s topmast shroud and parted it, which was followed by the cracking of the topmast, though it did not come down on deck. *Valkyrie III* sailed on and over the line, followed, according to contemporary accounts, by cries of 'Shame' and 'Outrageous', without waiting to see the damage she had done, or for that matter hoisting a protest flag. The first was a moral lapse which was never forgiven her; the latter was a technical error – if she believed she had a case – unjustified by Dunraven's subsequent

Lord Dunraven's *Valkyrie III*, the subject of an America's Cup race dispute which reached the level of an international episode.

weak explanation: 'I considered *Defender* responsible for the foul, and I ought perhaps to have protested. But I thought it possible the foul at the last moment was accidental, and I refrained from protesting.' The more one analyses that statement the more odd its implications seem.

In *Defender* a jury stay was rigged to keep the sagging topmast aloft, the jib topsail lowered to relieve the strain. A main topsail of some kind, it seems, was kept aloft – not presumably the jackyarder which was up at the time of the accident – and at some risk later in the race a small jib topsail was hoisted. Before *Valkyrie III* crossed the finishing-line *Defender* was pulling up on her from astern.

After the protest case had been decided in *Defender*'s favour, Iselin, managing owner of *Defender*, immediately offered to re-sail the race. This was the second chance that Dunraven had been offered for a re-sail. The first had been made by the Committee before the hearing of the protest. In either case it is creditable to Dunraven's sense of fairness to have rejected the offers. But he did so in ungracious terms, and made it clear that he considered the verdict to be mistaken.

There followed a day of busy note-passing between Lord Dunraven and the Race Committee. Dunraven's communications had a tendency to go astray. His first letter, for example, to Iselin refusing the offer of a re-sail was mislaid in transit – it would have arrived across the water quicker had he posted it – and had to be followed by another, equally unappreciative of the offer. Then he wrote to the Committee saying that the decline to race again unless the course was kept clear of 'steamers and tugboats . . .' On the evening of that day – with the third race due on the morrow – he had a meeting with the Committee, which agreed not to start the race unless the line was clear of interference.

After the meeting Dunraven sent another letter saying that the Committee must also agree to declare the race void if the boats were interfered with during the course, which was something the Committee had specifically covered at the meeting explaining that they had not the power to do so. This letter also suffered from the defective pigeon service and did not reach the Committee until it was too late to make further arrangements before the time of the start.

So, as we have seen, Dunraven sailed over the line and promptly retired. Soon afterwards he went home, and a typical newspaper comment said that he was easily the most unpopular Englishman ever to leave the country.

Valkyrie III and *Defender* passed quickly out of life, never racing again. Meanwhile Lord Dunraven brightened the autumn of his return to England with some widely publicized announcements. He wrote an article in *The Field* in which he stated without equivocation that *Defender* had been ballasted deeper between her official measurement and the first race. The fact that the yacht had been remeasured after the first race and no difference found since the earlier measurement proved nothing, he claimed, since no official observers had been living on board the two yachts between the measurements. He repeated all this in a well-reported speech. He was a not inconsiderable public and political figure.

He was a man also of intelligence, talent and charm. He had initial ambitions as a musician, and I have it on authority I cannot judge that he was a first-rate violinist. I need no authority to convince me he was a first-rate technical writer. Apart from being correspondent for the then *Daily Telegraph* during the Franco–Prussian war, he wrote a three-volume work entitled *Self-instruction in the Practice and Theory of Navigation*; and examining it for the purposes of this chapter, I felt what an example in clarity it was for much of today's technical writing on yachts. Dunraven held a Master's Certificate and he earned by his nautical experience and knowledge, not to mention his enthusiasm, his position as the best-known yachtsman of the day. The enthusiasm remained into old age, and when his quarrel with the New York Yacht Club had become an old, unhappy, far-off thing, he built *Sona* the first diesel yacht.

He was not only a clever man; he was far-seeing, ahead of his time in all his wide-ranging interests, and could be both wise and generous in his attitudes. In the troubled politics of Ireland where he was a considerable land-owner, he was a conciliator at a time when there were few and many were needed. In politics he had a liability to support the immediately losing side, but the side that time ultimately proved to have been the right one. He was fundamentally an artist in temperament, which brings vision but usually not a good judgement.

Essentially, I believe, it was Dunraven's vivid imagination, his lack of judgement in small, immediate matters, fired by a temper always easily rushed to the boil, that managed to overcome his native Irish charm and even his very keen sense of humour and led him to behave with some absurdity for several months in 1895, when he became a spider in the middle of a web of what can only seem most unpleasant gamesmanship.

Let us return to the America's Cup scene at Sandy Hook just before the start of the first race, and look at events through the eyes of Lord Dunraven. *Defender* lay not far off the *City of Bridgeport*, *Valkyrie III*'s tender, and Dunraven's attention was drawn to the fact that the American yacht seemed to be floating appreciably deeper in the water than she had been yesterday when she was measured. Lord Dunraven later claimed that not only his own crew but Americans who were with him in the *City of Bridgeport* were in agreement about his impression. Closer examination revealed that the discharge outlet of *Defender*'s bilge pump, which yesterday had been at water-level, was out of sight. Also, a seam of her bronze topside plating, which formerly had been above the water amidships, was now immersed. Another detail was that the lower bolt of the bobstay was nearer the water than hitherto; but Dunraven agreed that this might be caused by a trivial alteration in fore and aft trim. All on board *Valkyrie III*'s tender confirmed these observations; and there was a further sinister fact that *Defender*'s crew, who slept on board, had been heard noisily at work from sunset into the early hours of the morning. Loading ballast?

Such was the evidence that confronted Dunraven as the morning grew older and the time to set off for the starting-line became imminent. How hard is it to recapture the atmosphere of tense past moments. The facts alone cannot rekindle the embers. The air that morning must have been electric with pre-race tension, nerves strung with anticipation, the steadiness of judgement rocked by unremitting blasts of pepped-up publicity. Furthermore, we have it from an American source of information (Lawson's massive *History of the America's Cup*, printed for private circulation in 1902) that American friends of Dunraven who were unfriendly to Iselin had been spreading gossip that the latter had behaved questionably over the ballasting of a previous yacht that he had

The spectator fleet rushes after the yachts and adds fuel to the controversy by getting in the way of *Valkyrie III*. When the subsequent row was in full blaze, and while there was also some diplomatic tension between the U.S.A. and Britain, an English yachtsman cabled to an American: 'If we send our fleet to bombard New York will you be sure that your spectator fleet does not get in our way.'

controlled. This is the kind of innuendo that floats too easily around closed, small, highly publicized societies, in which good nature and healthy scepticism are usually the first casualties.

By nature imaginative and now made suspicious, after living for weeks in a hot-house world, the observations of that morning were enough to convince Dunraven that he was being cheated. The atmosphere and the urgency of the moment could do nothing to cast a saner light revealing the moral preposterousness of his idea or its practical absurdity.

So before the race was sailed, Lord Dunraven asked the American observer in *Valkyrie III* to request the Committee to have the yachts remeasured after the race. According to Dunraven he also asked that immediately after the race a member of the Committee or other reliable witnesses should be put on board each yacht until remeasurement was complete. The Committee subsequently denied that this request ever reached them. At any rate, nothing in this way was done. When the yachts were remeasured Dunraven appeared to accept the situation, and it was not until his return to England that he publicized the fact that in his opinion the remeasurement meant nothing. This is one of the peculiar inconsistencies in his behaviour.

Dunraven's state of mind during the next days makes a curious study in the collapse of the sporting ethic under the stress of overpublicized, overglamorized, overfinanced international competition. He believed sincerely that he had been cheated in the first race, and hoodwinked over the remeasurement. Convinced as he was, true sportsmanship would have led him to go through the form of completing the series of races with the minimum of fuss and thereafter washing his hands of such a travesty of a sport. Alternatively, if less ideally, he might have quietly withdrawn from the competition immediately, with some cryptic excuse, and thereafter kept quiet.

Instead, as we have seen, he continued the series in the most flamboyantly aggressive manner, shifted his line of attack from illegal ballasting to the crowding of the course, which he knew, and in more rational moments agreed, was a problem of major yacht racing that had at that time not been solved either in England or the U.S.A., and which he knew the Race Committee had not the powers to deal with; then, in a way to assure the utmost publicity and ill-feeling, withdrew from the series and the situation, only to stir and heat the pot of ill will yet more furiously by reverting to his original and much more serious charge of cheating in a press and public-speaking campaign.

Thereupon the pot finally boiled over. The New York Yacht Club established a judicial Committee of Inquiry, on which amongst others sat Pierpont Morgan, Rear Admiral Alfred Mahan, U.S.N., the internationally known authority on naval warfare whose reputation was high in England, and Edward Phelps, later American ambassador in London. An important technical witness was Nathaniel Herreshoff, designer of *Defender*, and Captain Haff, the yacht's professional skipper. Lord Dunraven appeared before the Inquiry and was represented by a Q.C., and Iselin and the Committee had their counsels.

The main impression left by the findings of the Inquiry was their revelation of the essential frivolity of Dunraven's charges and behaviour. The evidence on which he had based the accusation of reballasting was soon in tatters. It was shown by a U.S. Navy Constructor, called as an expert witness, that the bilge pump outlet, when the boat was upright in smooth water and ballasted as she was measured, was, in fact, a little below the waterline, but that a small angle of heel, such as might result from the shifting of the main boom to one side, or many

Thirty seconds before the foul, which occurred at the start, when *Valkyrie III*'s mainboom swept the deck of *Defender*, seen astern here with the Committee boat to the left. Another photograph taken thirty-five seconds later shows *Defender*'s topmast bending acutely to port after the parting of her starboard topmast shroud, which had been fouled by the end of *Valkyrie*'s boom.

other causes, would be enough to lift it temporarily out of the water. The movement of the thirty-man crew from amidships to their quarters in the fo'c's'le was enough to account for the slight change in fore and aft trim revealed by the lower bolt of the bobstay. The 'noises in the night' were easily explained by the fact that a couple of tons of ballast that for reasons well known to everyone had been taken on board just before her first measurement and simply laid on the cabin sole, were being fitted into the bilge, and this entailed cutting the pigs in two. Dunraven's facts, on which he had based his serious accusation, could be disposed of by the simplest explanations.

But the more effective revelation of the absurdity of his charges lay in the fact that to have produced the change of floatation in *Defender* believed by Dunraven to have occurred would have required 13 tons of ballast. To substantiate the charges, explicity made by the latter before the Inquiry, would have required: (1) the removal of this weight prior to her first measurement; (2) its replacement immediately afterwards; (3) its removal after the race for the remeasurement. Each would have been an operation of magnitude undertaken with the connivance of the whole crew of thirty, organized by mates and the skipper, all of whom would have had to be trusted never to divulge the secret, drunk or sober. The affair melted into absurdity.

But more interesting, I think, from our immediate point of view, was the biting observation made by the American counsel for the Race Committee to Lord Dunraven, concerning the period after the latter had assured himself that he was being cheated and while he kept on board as official observer in *Valkyrie III*, the American representative who must have been party to the fraud: 'So you were willing to sail with a fraudulent party, with a fraudulent rival, under a Cup Committee who refused you any opportunity to prove the fraud. . . .' This surely was frivolity revealed?

The findings of the Committee, which nobody on either side of the Atlantic questioned, of course absolved Iselin and all members of *Defender*'s crew management and paid crew.

So ended the Dunraven affair, in which we seem to rise on a crescendo of unreasonableness to a climax of misplaced, misapplied seriousness of a kind that has no place in sport. While the affair was at its hottest there happened also to be a period of diplomatic strain between Great Britain and the U.S.A., which this yachting occasion did its little to augment. The sanest thing about the business was a cable sent at the time from an English to an American yachtsman: 'If we come and bombard New York will you see that your pleasure steamers do not get in the way.'

And that, oddly enough, is the kind of message that at other times Dunraven himself would have delighted in producing.

	No. of yachts	Tonnage
United Kingdom	2,959	53,025
Empire	311	4,563
Germany and Austria	599	8,371
France	363	6,300
Norway and Sweden	300	4,369
Belgium and Holland	191	2,543
Denmark	107	1,755
Italy	76	1,726

We have now to consider some matters of international organization which may appear deficient in the sea's tang and vigour, but which were of crucial importance to the future development of yachting. It is hardly an exaggeration to say that during 1906–7 there was ushered in the modern era, with a new framework of organization which has been evolving ever since whilst retaining its original form.

The table, left, gives a clear idea, without pretending to exact numerical accuracy, of the European distribution of sailing yachts in 1906. The United States had now become the greatest yachting nation, with a total tonnage exceeding that of Britain and what was then the Empire by some twenty-two per cent, while Britain alone possessed not much less than double the tonnage of the rest of

Europe. It will be noticed that the second largest yacht-owning nation in Europe was Germany and Austria combined, though the empire of the Habsburgs did not contribute much to the total. The last German Emperor had now been on his throne for nearly twenty vivid years, for ten of which his programme of naval expansion had been growing in crescendo, and with it efforts to make the German people nautically aware. Brooke Heckstall-Smith wrote of the Kaiser William II that he 'told me all about his yachts and how he came to take the sport so keenly. It was not merely that he was personally fond of racing; he said he wished to encourage yachting in Germany and maritime sport to induce the

The Kaiser's schooner, *Meteor IV*.

German people to be interested in maritime affairs, and so gradually to work up their enthusiasm for his navy. He was perfectly frank about it, and at the Kiel regatta in 1912 . . . he told me it had taken him nearly twenty-five years – since the time of *Meteor I* – to train an all-German crew to be as good as a British crew.'

It will also be seen from the table that Norway, Sweden and Denmark combined, with a fraction of Germany's and Austria's population, had three-quarters of their tonnage of yachts, and in the U.S.A. the reputation of Scandinavian professional crews was unrivalled. In Germany yachting was a royal sport, in Scandinavia a people's sport. The differing social conditions of yachting in the countries of continental Europe may be discerned from the figures of the table. Italy with only three-quarters the number of Denmark's yachts had much the same tonnage as Denmark. Despite the massive preponderance of Britain's yachting fleet over that of all the rest of Europe, the shape of things to come was prefigured most closely in the Scandinavian nations. But it was also inevitable that Britain should take the European lead in the international organization of the sport.

Four years of splendid Big Class racing was followed by a decline which persisted until the last phase of this sport during the 1920s and 1930s. Many big racing yachts appeared, some superior to those of 1893, but they failed to coalesce into a hard racing, hotly competitive Big Class able to attract the attention of the public as well as yachtsmen.

Competition led to a deterioration in the type of yacht produced by the unsafeguarded length-and-sail-area rule, which briefly for a few years had raised the beauty of *Britannia* and her immediate contemporaries – 'that model which was so sweetly and so harmoniously blended in all its elements', as James Meikle, who had seen *Britannia* and *Valkyrie II* under construction, described it. There was nothing in the rule to prevent designers producing yachts that were shallower in body, lighter in displacement and with more depth of keel than those of preceding years. And yacht architects were avidly grasping the fruit from the tree of design knowledge during this time. In even the largest classes, in which experiment tends to be more restrained than amongst the smaller craft, yachts tended towards the skimming dish type. While Big Class racing under the tonnage rules had run to seed with yachts of extreme narrowness and deepness, under the length and sail area rule it was the reverse extremes of beam and shallowness that occurred.

A new rule was therefore introduced in 1896 followed by a modification of it in 1906, the First and Second Linear Rules respectively, which departed from the simplicity of the length and sail area rule and encouraged a more full bodied type of yacht. The rules sought, in fact, a return to what was already known as the '*Britannia* ideal', the yachts of the 1893 season which the length and sail area rule had produced before yacht architects had discovered its inherent weakness.

In 1906, then, British yacht racing was governed by the Yacht Racing Association's racing rules, as accepted by all clubs in 1882 and, for the measurement of yachts, by the second linear rule which was due for revision in 1907. At this time not only Britain and the U.S.A., but all the nations of Europe, were using racing and measurement rules subtly different from one another. Brooke Heckstall-Smith, Secretary of the Y.R.A. and destined to remain in office until his death in 1944, had for a few years been working out the idea of bringing the countries of Europe together for the adoption of a common measurement rule for racing yachts. In 1906, at the instigation of the Y.R.A., a conference was held in London the primary purpose of which was to formulate an internationally

Opposite

Puritan leading *Genesta* in the last leg of the final race for the America's Cup in September 1885.

recognized system of measurement. This, a further adaptation of the British linear rule, became the First International Rule of measurement in yachting history.

The nations assembled in London made the further considerable historical step by forming the International Yacht Racing Union, the members of which initially were Britain, France, Germany, Norway, Sweden, Denmark, Holland, Belgium, Italy, Switzerland, Austria–Hungary, Finland and Spain. The greatest yachting nation – the U.S.A. – did not become a member, nor at this time did it have any national authority, which is comprehensible when the immense length of its coastline is considered compared with the relatively tightly packed yachting communities of the European nations. When yachting interests of the Great Republic addressed the Old World it did so usually with the voice of the New York Yacht Club.

At Paris in the following year the young I.Y.R.U., its initials yet unfamiliar in a world whose instincts were certainly not yet internationally inclined, set itself the further task of adopting uniform racing rules, to match the now uniform measurement, or rating, rules. This was successfully accomplished. Essentially the rules adopted were those of the Royal Yachting Association, but with slight modifications in the section dealing with the management of races. Contemporary discussion on this subject throw some amusing sidelights on the differing national attitudes to the social aspects of yachting. In some European countries the convivial feasts following a race, the speeches and the presentation of the cups, may have been no less important than the hours spent out racing round the marks. It was thus essential for the results of a race to be known immediately after the last finishing guns had been fired; which entailed that a yacht's rating certificate, from which her time allowance or even eligibility for the race was determined, had to be up to date before she started. Other yachtsmen, more intent on achieving as near perfection as possible in the performance of their craft rather than on ceremonies after the race, wished to be free to make adjustments in trim up to the last moment practicable; when it might be too late for the measurer to come on board to check the rating, thus making it necessary for the rating to be confirmed after the race, and delaying the publication of results for a matter perhaps of days. Such national differences in attitude to racing were catered for by leaving certain matters in the international rules for the management of racing to the discretion of the national authorities.

A difference in national attitudes to yacht racing, though not directly concerned with rules, was emerging during the 1890s between the two biggest yachting nations, Britain and the U.S.A., who together at this time possessed some five times as many yachts as the rest of the world combined. Compared with most European yachtsmen the British leaned more firmly towards the maritime rather than the social aspects of the affair; though in this connection it should be remembered that in the pre-1914 Cowes Weeks yachting put on a purely social parade that Kiel Week tried vainly to emulate and nowhere else attempted to rival. But always there was enough wet salt in the waters of the Solent to assure that the yachting should not decline into mere blandness. But . . . what was it . . . was there perhaps lack of the more biting edge of competitiveness in British yacht racing? Let Brooke Heckstall-Smith speak again, for he was a closer observer for a longer time of the grand yachting scene than anyone:

'I am, however, only stating a fact when I say that the majority of British "yachtsmen" in the highest sense of the term do not attach much importance to the "cups". They do not take any particular trouble to win them, and have to be worried or "requested" by some club or "gingered up" to enter for them.

Above

Valkyrie II loses a spinnaker in the America's Cup race against *Vigilant* on 13 October 1893. Painting by Barlow Moore.

Below

A painting in the New York Yacht Club by James Butterworth showing the schooner *Columbia* with the main topsail and jib topsail being lowered.

'This may be a good thing for our yachting, or it may not, but it is true.

'When the "Olympic Yacht Races" were held in Amsterdam in 1928, every foreign country, including America, sent their best yachts and "teams" of amateurs as crew to sail in these "championships". On behalf of Great Britain one good sportsman came forward and did his best with quite a second rate boat in one of the matches, whilst in another event not a single British yachtsman could be found to compete or take the slightest interest in the affair. This has always been the way of English yachting people. It was the same thirty-six years ago as it is today. "They do not give a damn for cups." That is to say, for these special championship cups. The greatest of all is the "America's Cup".

'This is regarded in America as – well, I really do not know what. The New York Yacht Club look upon themselves as custodian of the America's Cup; it is regarded as something apart and beyond all other sporting events by American yachtsmen. But English yachtsmen do not look upon it in that light – that is to say, not the practical yachtsmen who know all about yachting. They regard it nowadays as "Lipton's affair".'

The above is not to be agreed or disagreed with but heard as the voice of Victorian sporting England. The voice is robust enough but is not strong in any competitive awareness; it speaks with the undertones of village green cricket, of hunting fields where a good gallop and fences taken cleanly, rather than the cut and thrust of pure rivalry, are what makes the best day's sport. To understand a sport so international as yacht racing, national attitudes with roots deep in the past, must be appreciated. The above observations are able to account for

Britannia and *Ailsa* (up to weather) racing in the Solent. *Ailsa* came out in 1895 and was one of *Britannia*'s principal rivals. Painting by J. Edmunds.

much in the lively history of Anglo–American yachting: and in the following pages we are able to watch the operation of national attitudes in the competition between all sizes of yacht during the twentieth century. The thoroughness, the dedicated approach, unsparing of energy, shown by the yachtsmen of the U.S.A. raised race sailing techniques to a new level of sophistication, which others followed with varying degrees of dilatoriness.

When the last season of the old century opened Big Class racing was firmly on the rocks, despite the fair course that appeared to have been set for it six years earlier. And thus it remained until, in the transformed world of the 1920s, the last Big Class in the history of yachting was racing between the shadows of two wars.

In 1897 *Britannia* and *Ailsa* had spent much of the summer racing in the Mediterranean, a perfectly matched pair, returning only for Cowes Week, while *Satanita* remained laid up in her mud berth. By 1899 the Big Class was defunct as a well organized racing entity. During the first half-dozen summers of the new century there was little change. Handicap racing prevailed among the big yachts, with the disadvantage that this was a haphazard affair, largely controlled by the judgement of the handicapper; a man who while doing his best would find himself the focus of some barely concealed odium. Some of this came from the German Emperor, which may have brought sympathy rather than the reverse to the handicapper, but the fact was that the system of handicapping by performance was inimical to serious yacht racing and quite unsuitable for application to the finest racing yachts of the day. There were attempts to revive Big Class racing, supported by the cutter *Nyria*, which after the war was to become the first of her large type to carry the yet experimental Bermudian rig. So affairs stood in 1906.

So far as the Big Class was concerned, the twentieth century, in which this kind of yacht racing was to become extinct before much more than a third of the century had passed, opened as an augury of the future. Between 1900 and 1914 a number of large racing cutters appeared for the first time or remained in commission from the flourishing years that had just passed; but there was no active exciting small group of the largest racing yachts, similar in size but interestingly different in character, which could engage in close competition in the round-the-coast regattas, capturing the public imagination and setting a crown that could dazzle on the yachting scene. There is a particular quality in Big Class yacht racing that no other can have. The character of the years 1900–14 was expressed in July 1912, by the Editor of the *Yachting Monthly* in a survey of the Big Class yachting scene which concluded: 'Yacht racing does not appeal to the public, even the beautiful *Shamrock* and *White Heather* do not excite as much interest as a Punch and Judy show.' The trouble was that potential owners of Big Class yachts seemed to feel like the public.

7
The small craft

One aspect of Solent yacht racing in the 1880s was described thus:

'. . . a new sport has been born – the racing of small yachts . . . the very perfection to which racing has been brought (in large yachts) tells in the same direction, because few men can afford to build large racers year by year to replace those that are outclassed. Yacht clubs have increased both in numbers and wealth, and the executives find that racing brings grist to the mill and repays the cost and the trouble. This specially applies to small yacht races . . . Owners are not slow to avail themselves of the sport that offers, which on trial proves to possess many advantages over large yacht racing.'

This new development in the exclusive Solent waters, where hitherto large yachts had dominated the scene, was at once the precursor of later sailing days in the twentieth century and also the then latest aspect of a maritime activity that was by no means new in the 1880s, but had been proceeding at numerous places along many coasts without publicity and with no annals to record it. We should briefly glance at the earlier years.

During the years leading up to the 1870s there was beginning an activity, in harbour, estuary and up-river waters, which was to expand during the following century into a considerably social movement in taste – the sailing and racing of small boats for pleasure.

What was to become a century later thousands of different classes of racing dinghies, handled by millions of people belonging to the twentieth century's egalitarian societies, began beyond the focus of history's records. Nothing is now known of those who first took their own small boats out sailing for fun on the waters of the Upper Thames, on the Clyde, on the Delaware river and New York's Hudson and East rivers, and in countless bays and harbours scattered along the coastlines. Sailing boats for pleasure no doubt grew initially as an extension of using boats under oar or sail for utilitarian purposes of transport, in those localities where it was easier to cross a stretch of water under oar or sail than to travel round it by bad roads and awkward bridges on foot, or by horse or pony gig. In such places all people tend to become amphibious. To take to the water not only for convenience but relaxation would have been natural in numerous areas where the yet rare yachts had never been seen; where they were the faint far echo of a grander, richer world.

Hunt's Magazine, where records of early yachting are preserved in a form quaint to modern tastes, recorded nothing of this small boat sailing, and not until the 1870s and 1880s does history hear of it.

Cat rigged boats were being sailed about New York and Boston in the early 1850s; also in the area of Cape Cod. In Britain Ingrid Holford has recorded, in her valuable booklet on *A Century of Sailing on the Thames* how a small body of men 'of that class of person who cannot bear to see a stretch of water without getting afloat' were sailing on the Thames at Surbiton in the late 1860s, and in 1870 the Thames Sailing Club was founded. A century later, at the International

Boat Show in London in January 1970, this club exhibited one of the old Thames raters, still preserved with the aid of plastic sheathing. In 1871 the Rear Commodore of this new up-river club flew its burgee in Russia, on a visit to St Petersburg in his 30-ton cutter *Thalatta*, an international gesture of recognition towards the new sport of small boat racing. Meanwhile, at home the club's Commodore was organizing the 1871 regatta on the Thames, which Ingrid Holford has described thus:

'The gig race was open to all *bona fide* rowing boats carrying one sail only and the property of gentlemen amateurs, under the auspices of, and for a handsome cup given by, the Thames Sailing Club. More than 80 boats proposed to enter and 57 actually entered and started. The race excited much interest, and river and banks alike were covered with spectators, ladies mustering in force. Instantaneous hoisting of sails of a score of boats moored in line across the river had a very pleasing appearance. The arrangements for starting were especially good. . . . The committee did their part to perfection, but nature was not kind in the matter of wind, which by the time the final heat came to be sailed had died away altogether. Two boats were disqualified for attempts to row with the hands. Much amusement was caused by the "dodge". One ingenious gentleman, tired of drifting, commenced working his rudder somewhat a la screwpropeller, with such effect that others were obliged to follow suit – with even greater success.'

At this same period the small boats were out on very different sailing waters far to the north, on the Firth of Clyde, described thus by a contemporary:

'. . . during the summer months, in the bright evenings, from every coast village may be seen a fleet of little vessels flitting along the shore in smooth water, and lying over to the land wind, which, in good weather, rises as the sun sets. Many of these boats are racing craft, and as each principal watering place has its club, there is no lack of sport on Saturday afternoons, there being always one, sometimes two or three matches for the little ships.'

Those boats that flecked the Clyde waters on mid-Victorian summer evenings were evolved from simple fishing skiffs. They became more powerful and better canvassed. Clinker built open boats with transom sterns and plumb ends, their era is most nakedly revealed to us in the fact that just above the turn of the bilge over the middle length there was a rack to take the movable ballast of shot bags (though not all clubs allowed movable ballast, when it had to go into the bottom) and crews were generally limited to three in the 17 ft and 19 ft boats to minimize the effect of live ballast. This brings us head-on with two now utterly discarded but once conventional Anglo–Saxon attitudes towards small open boats: (i) there was nothing objectionable to their having fixed ballast; (ii) live ballast was slightly disreputable. Today these two attitudes are precisely reversed.

The boats were usually adapted for two rigs, One was single standing lug; the other had a smaller lug on a mast stepped further aft with a jib on a pretty long bowsprit; both rigs were of gigantic area to modern eyes, and with willowly swaying yards aloft. The combination of the rig and the ballasted hull is alarming to modern ideas of safety; but accidents were rare and all clubs enforced one rule that seems familiar to a century later; boats had to carry life-saving equipment to float every person on board. The best of these boats were considered 'very expensive' by a contemporary of them in 1878; a 19 ft boat then cost £75, complete down to three lifebelts at 75 pence each; and adjusting to the scale of today's inflated currency, we must agree.

Over on the Irish coast the oldest one-design class in Europe was introduced in 1887. It is significant, in the light of the future developments in dinghies, that the Water Wags of Dublin Bay were unballasted centreboard sailing boats.

They were 14 ft 3 in in length, 5 ft 3 in in beam, and not only became widely known in their early years, when the character of their design was a novelty, but enjoyed a continuing fame, marching on into the revolutionary twentieth century, and surviving today unembarrassed by their years.

The Water Wag story is similar to that of many dinghy classes since. They were intended to be a simple inexpensive type of boat; but the construction was at first inadequately covered by the drawings and specification, with the result that the Water Wags' popularity soon caused the price of the boats to rise from £15 to £28; an increase of ninety per cent in a few years, which indicates that racing

dinghy developments could be as brisk in the 1880s as in the 1970s. Twenty-eight pounds was a considerable sum to pay in the 1880s for a 14 ft clinker-built open boat with a standing lug rig. So the Wags were redesigned to a tighter specification; the rig became a well proportioned gunter sloop; planking had to be of the then cheap yellow pine instead of the lighter cedar that had been used in some of the earlier boats; and refinements of finish, such as fining the outside lands of the planks or bevelling the corners of the timbers, were outlawed. A modified design and specification was introduced in 1900, and Wags appeared in clubs by the Indian Ocean as well as in Dublin Bay.

A 'Hiker' race on the Delaware river *c.* 1875.

In the 1870s Herreshoff was making experiments with catamarans. These are of peculiar interest today when multi-hulls are receiving so much attention for inshore and offshore racing and even ocean voyaging. It may be said that the general acceptance of the type, which has been a feature of the 1950s and 1960s, and the further remarkable extension of their use offshore, which recent experience has shown to be justified, owes little to Herreshoff, who after brief enthusiasm appears to have lost interest in the type.

It is in connection with catamarans that we find one of Herreshoff's rare communications to the press, appearing in the *New York Herald* dated 10 April 1877, when he was twenty-nine years old. A cutting of the lively and lucid letter survives in the New York Yacht Club:

'In the fall of 1875 I was thinking and thinking how to get speed out of single hulled boats, of the kind in common use. To get great speed, thought I, one must have great power. To have great power one must have great sail, and you must have something to hold it up, and that something must be large and wide, and have a large sectional surface, and also a great deal of frictional surface. These properties in a hull to give stability are not compatible with attaining great speed. Indeed, the more one tries to make a stiff, stable hull the less speed will be attained, even if corresponding additions are made to the sail. So, then, there are two important principles of speed which constantly work against each other. If we increase the power to get more speed we must increase the stability of the hull correspondingly. An increased hull has more resistance, both from sectional area and surface friction. So what we would gain one way we lose in the other. . . .

'. . . A wide boat cannot have great speed, apply what power you will to her, so the next thing that is to be done is to decrease the sectional area and the frictional area. I thought I would raise the keel and the center line of the boat, and make the bilges project downward and outward from it – such a thing as a Dutchman might build . . . I kept on following this principle, getting the keel higher and higher, when, lo and behold! There was a double boat! Nothing else to be done but take a saw and split her in two, spread it apart a little way and cover all with a deck and there you are. That was the rough road which I travelled, and having arrived this far I abandoned my ill-shaped hulls and in their place substituted two long, narrow, very light boats and connected them at the bow, stern, and middle.

'The extreme length of the hull is 24 ft 10 in . . . the width of each hull is 20 in in the widest place and 18 in on the waterline.

'Sailing in a catamaran is an entirely new sensation, and it has everything in a recreation to recommend it, safety being one of its chief attributes. There is no shifting ballast, no hanging on with tooth and sail up to windward. I am sure that a half day's sail in *Amaryllis* would spoil anyone for the old fashioned sailing. . . .'

Herreshoff, travelling the 'rough road' towards the double hull concept, clearly was no more influenced by Polynesian boats than Polynesian grass skirts. His approach to the matter was that of an engineer and a hydrodynamicist. It will be noticed that, with his ideas conditioned by the traditional beamy type of American yacht whose stability was primarily due to form rather than ballast, he regarded the catamaran as a means of achieving stability without excessive beam in a single hull, whereas the modern approach has been to consider it primarily as the means of dispensing with excessive ballast. But his and the modern approach are in principle identical – the multi-hull boat provides the stability without the power of hull, due mainly either to beam or ballast, the effect of which is to offset any gain resulting from the greater sail area that may be carried.

The catamarans that Herreshoff produced, of which *Amaryllis* was the first, bore no close relationship to the rudimentary craft of the South Seas, nor to the multi-hulls of today. The hulls were linked by a flexible system of ball-and-socket joints allowing each hull to adjust itself to the momentary water levels without producing heel, and the car between the hulls containing the crew remained sensibly level. There was no question of sailing with the weather hull just lifted, and a speed advantage was thus lost. On Herreshoff's evidence she regularly made 15–18 knots – other claims were higher – and she sailed rings round a fleet of ninety conventional yachts during a regatta in New York harbour. Some Herreshoff catamarans were used for modest cruising. Their speed on a reach caused them to be described as 'aquatic marvels', though their performance to windward was not impressive; for a time there was limited enthusiasm for them.

But Herreshoff returned to the 'old fashioned sailing' and in later years believed that for average speeds over any distance of seaway he could achieve better results with single hulled craft.

Before small boat racing was of significance in yachting, the fishermen of Itchen village, by Southampton Water, had been racing their boats, which became one of the best known types of local working craft on the coast. The hamlet by the shore of the Itchen river is now submerged in macadamized and neon-lit Woolston, a part of the greater Southampton, and the breed of the Itchen ferrymen has gone too. As yachting became fashionable, Itchen village became well known for providing yacht hands, fishing in the winter, yachting in the summer, and from their number emerged some of the great yacht-racing skippers.

The Itchen ferry boats, often known as punts, were wholesome little cutters ranging in length between 22 ft and 28 ft overall with generous beam and a relatively shallow body. While the fishermen's racing of their craft did nothing to destroy their economical character and sturdy toughness, when yachtsmen adopted this attractive type for racing it was inevitable that money and enthusiasm should have gone far to ruining the punts' virtues. Waterline length alone was measured for racing. The simple transom and snubbed bow of the true punt were drawn out into long overhangs. The shallow hull of the type was deepened and heavily ballasted with lead internally to carry an immense spread of sail. There was one boat of the class 30 ft long overall which carried a spinnaker pole 50 ft long jointed like a fishing rod so that it might be stowed on board. The virtuous punts had been transformed into strumpets of racing machines.

Other small craft in the Solent were divided into classes for racing by overall length, and classes with upper limits to length of 21 ft, 25 ft, and 30 ft were introduced. It was during this period, in the mid-1870s, that R. S. S. Baden-Powell, later a distinguished general and founder of the Boy Scout movement, had built to his own design a 5-tonner 26 ft in length and 7 ft beam. It is believed that this was the first yacht of so small a size that apart from racing, was sailed from port to port on the south coast between races, her owner and friends living on board.

The oldest clubs on the Solent, the Royal Yacht Squadron and the Royal Southern Yacht Club at Southampton, were not at first active in promoting small boat racing. Initially support came mainly from the Royal Southampton and the Royal Portsmouth Corinthian Clubs, the latter surviving until after World War II, when it was merged with the Royal Albert Yacht Club of Southsea. The Royal Southampton is still active. Other clubs followed the lead – the Royal Albert, Royal London and especially the Royal Victoria of Ryde. The Squadron dragged its feet more than the rest, and it almost, for a time, earned the couplet often applied to it:

Nothing less than 40 T
May ever race with our burgee

The classes measured by length were replaced by the 'rater' classes built to what was basically the same Yacht Racing Association rule of 1887 that governed the design of the biggest class, *Britannia* and her noble consorts. It is always so that adventurous design emerges first amongst the small craft; and the smaller raters of the 1880s and 1890s brought to the art of race sailing something comparable to the then 'modern' art that was so shocking the world brought up to classical painting. But few people who sailed ever came to grips with Raoul Dufy's later representations of yachts.

Initially there were classes of 10-, 5- and 2½-Raters established in 1888. Since rating was equal fundamentally to the product of length and sail area, the size of the boats in each class might vary widely, some boats taking most of the rating in big sail area, others in more length and less sail. There was much discussion amongst the more technically equipped about the future of design in these classes. Scientific naval architecture, at least in its application to sailing yachts, having not yet lost the bloom of innocence, it was not yet evident that in each class the boats would get larger and the sail area less. Tradition and common sense suggested the reverse. The brute-force principle of sailing had all the traditional strength of its years of application. In 1881 Dixon Kemp was lending his great authority to the idea that 5-Raters might grow from 20 ft to 30 ft in waterline length; by 1892 they had become 34 ft long on the waterline. The 2½-Raters he believed might lengthen from 16 ft to 20 ft. By 1892 they had become 28 ft on the water.

Two more smaller rating classes were formed, of one rating and one and one half. The later King George V when Duke of York had a boat in the latter class, the *White Rose*, which was racing in the last year of the class in 1895. Here developments in design were even more rapid. The length of 2½-Raters increased from some 17 ft to 28 ft, while in the 1-Rater class it doubled from 10 to more than 20 ft. Science was continuing its *danse macabre* among the racing yachts. As the boats of each rating became larger, a class rating of only one-half was formed. The smaller rating class boats inevitably became expensive for their size for they were the yet most perfect small racing machines ever to have appeared. The merits of the various classes were much discussed, expense being regarded as not unimportant. The opinions of Lord Dunraven, who raced in the 5-Rater class, have an air that is at once both modern and thoroughly dated:

'Nothing can be more delightful than a 2½. It is the perfection of racing of its kind; but the absence of any accommodation below is a serious drawback under certain circumstances, especially to persons living at a distance.

1-Raters racing in the Solent. To windward is *Red Rover*, to leeward *White Rose*, belonging to the Duke of York, later King George V. In 1895, the year of this photograph, 1-Raters were about 15 ft 6 in on the waterline and 23 ft overall. Under the rater measurement rule hull size might be increased at the expense of reduced sail area.

'On a 5 you can change clothes, boil a kettle, and, on a pinch, sleep.'

He wrote thus in 1889, by which time a typical 5-Rater was some 48 ft overall, 34 ft on the waterline with a displacement tonnage of some 9 tons – by the standards of eighty years later a yacht in which might be found not a kettle and a pinch of sleep, but living quarters for a family or a tough racing crew of eight.

The 1880s were a time of exciting development in yacht design, and this found its most adventurous expression in the small classes, especially the 1- and ½-Raters. Yacht architects were still feeling their way through the new science that had descended upon them. Boats in these two classes might be centreboarders or have deep lead-loaded fin-keels such as Herreshoff had introduced in America – there was much discussion on the merits of the two types, which eventually was emphatically decided in favour of the keelboats. Before designers had appreciated the importance of length for speed, the boats had transom sterns and short forward overhangs; later, to gain unmeasured length, the overhangs reached out far beyond the static waterline, and the boats became extreme skimming dishes.

In rig, as in hull design, the little rating classes were reaching for the future. The huge sail area in proportion to their size converted even the ½-Raters – strictly comparable to today's Flying Fifteen class in length and displacement but carrying up to twenty-seven per cent more sail area – into three- and four-men boats instead of the two adequate today under the same Solent conditions. The rig in the early days was the gaff sloop with large mainsail without topsail, short luffed and long footed, and with a very small jib. But sometimes topsails, even jackyard topsails, were carried, as in the larger rating classes.

These clumsy and inefficient rigs were replaced by a kind of lugsail devised by the sailmaker Tom Ratsey, which swept the board in the rating classes up to the 5-Raters. In this rig the low peaked gaff with its jaws well up the mast was replaced by a longer yard slung from a halyard rather more than two-thirds of

A famous $\frac{1}{2}$-Rater, the smallest of the rater classes, was *Wee Winn* designed by Nathaniel Herreshoff for Miss Winnifred Sutton in 1892. She proved herself the decisive champion of the Solent in all weathers. *Wee Winn* was 23 ft 10 in overall and 15 ft on the waterline. Here she is shown in a stiff breeze and reefed well down.

the yard's length from the heel, the yard then forming a continuation of the mast, as nearly as possible in the same line. The mainsail was increased in height and approached the triangular in profile; but the yard, doubling the short mast for about half the latter's height, added weight aloft and clearly invited the appearance of what was to become the Bermudian rig, the yard eliminated, the mast taller, the mainsail a true triangle. This was to become the standard racing rig, but not for a few decades yet.

Late in the summer of 1892 a couple of fin-keel boats reached Britain from the Bristol yard of Herreshoff, one to cause a stir on the Clyde the other in the Solent. It may be recalled that this was the year when the fin-keel *Dilemma* and her brood, also from the Herreshoff Manufacturing Company, was disconcerting yachtsmen in Long Island Sound.

The Clyde boat was the $2\frac{1}{2}$-Rater *Wenanah*, designed to the English length and sail area rule, which gave full advantage to the short waterline and long overhangs of the Herreshoff *Dilemma* conception.

She became the Clyde champion, winning seventeen races and gaining a third and two second flags out of twenty races. Meanwhile the other Herreshoff boat, a $\frac{1}{2}$-Rater sailed by her owner Miss Winn Sutton – who called the boat *Wee Winn* – was even more convincingly victorious on the Solent, winning twenty first flags and a second in twenty-one races. It was noted with pleasure by a contemporary English observer that the two brilliant Herreshoff boats carried what he described as 'regular gaff mainsails', not the 'lugs for which there is a craze'. The latter were the above mentioned Ratsey type of lug.

1892 was also the year of *Dacia*, which carried such a lug, a 5-Rater produced by that rising star among yacht designers and builders in the south of England, Charles E. Nicholson. Yachting commentators were combining the names of Nicholson and Herreshoff, names which were to dominate yacht design and construction for a generation in two continents: 'This year was full of surprises in the $2\frac{1}{2}$-rating class, as indeed in all the small classes. These were mainly due to two gentlemen, Mr Nat Herreshoff of Rhode Island, U.S. and Mr C. Nicholson (jun) of Gosport, G.B.'

Despite the immediate success of *Dilemma* and her like, the fin-keel did not gain wide adoption, and indeed had to wait until after World War II to win wide acceptance.

The year 1895 deserves to be memorable in the yachting story, for it saw the first contest for the later celebrated Seawanhaka International Challenge Trophy for Small Yachts. It was the outcome of a challenge made by an English yachtsman, J. A. Brand of the London Minima Yacht Club, for a series of races in the English $\frac{1}{2}$-Rater class. The Seawanhaka Corinthian Yacht Club not only provided the silver cup but had to raise a fleet of boats, for the class was not sailed in America. Six new fin-keel and centreboard boats were built, and the selection fell on *Ethelwynn*, which became the first defender of the Seawanhaka. This little centreboarder, of some 15 ft on the waterline and with long overhangs, themselves totalling in length about half that of the waterline, was an augury of future Seawanhaka Cup racing, which has always encouraged the most advanced small racing machines of the day. *Ethelwynn* was Bermudian rigged, the first American yacht, it is believed, to have carried the jib-headed mainsail. She met and defeated the challenger, *Spruce IV*, in Oyster Bay, winning three out of five races, which were sailed just after the America's Cup contest of that year between *Valkyrie* and *Defender*.

In the following year Canada challenged for the cup. The defence raised a fleet of no less than twenty-eight boats from seventeen clubs from which to select the defending boat. Five of these belonged to the Seawanhaka club, including *El Heirie*, the boat ultimately selected. Her designer, Clinton Crane, was later to design a boat for the defence of the America's Cup, *Weetamoe*, one of the five yachts built for the defence in 1930. His was a unique achievement in that though trained as a naval architect, he spent most of his life in business, and in 1930 he was essentially an amateur.

In 1896 he was hardly old enough to have been regarded more than an amateur, being at the beginning of his professional studies. Crane's remarks on the trial races are significant: 'It is amazing now, looking back, to remember the interest which was taken in them. Day by day a group of reporters arrived representing every one of the New York papers, a man from Boston, and men representing each one of the yachting magazines. And in Boston the reports of the races were often on the front page.'

But *El Heirie*, victor from a fleet of twenty-eight, and in spite of having had her bottom polished by a piano finisher from the firm of Steinway, was beaten in three straight races by the Canadian challenger *Glencairn*, of the Royal St Lawrence Yacht Club.

The Seawanhaka Yacht Club challenged in the following year. The simple length and sail area measurement rule gave itself to violent experiments in design. The Seawanhaka rule that had been adopted for the contests was basically similar to the one in force in Britain, but the Americans brought to it a more violently experimental attitude. The fleet of boats that gathered for the trial races to select a defender in 1897 was described by the American yachting historian W. P. Stephens as 'the most grotesque collection of craft ever seen in civilized waters, the name "freak" being one of the mildest applied to them'. The boat selected was representative of the fleet, more than twice as long overall as her waterline length and so shaped that when heeled the whole of her length was immersed. Having a hollow along the length of the bottom she was essentially double hulled below the waterline, and efforts were made to disqualify her on the grounds that she was a catamaran and barred from racing with single hull yachts. But the claim was not allowed; and the boat swept to victory against the Canadian challenger without difficulty.

The extreme scow type of hull had entered yachting with these races. The fame of the cup spread, and with it an interest in small boat racing; it was yet surprising that international yachting competition should be contested in such small and relatively cheap craft. Able to race on small bodies of water, yacht racing in such craft spread to the western states, a trickle that was to grow into a flood, an inundation, in the years ahead.

The small raters and other such craft in the U.K. and U.S.A. were not poor men's compensations for the large yachts which made the superb backcloth against which the little ones raced. They provided the sport that was yet young of amateur helmsmanship, at the time when the joke against the big racing-yacht owner was still topical: 'Would you care to take the helm, Sir?' inquires the professional skipper. 'Thanks Captain, but I take nothing between meals.'

Some of the men, and women too, of the small racing classes did little else during those summers of leisure in that stable old world. Especially was this so in Europe. They may not have raced before breakfast or after tea in the modern offshore racing manner, but there was salt in their blood, as one feels it is not always in the blood of those who now polish their plastics dinghies, the better perhaps they may see to comb their hair in them. Both modes of sailing are

Dominion, which won the Seawanhaka cup in 1898, was an outstanding freak amongst a fleet of unusual craft. While 37 ft long overall, her waterline measurement was only 17 ft 6 in, and this was the length by which she was rated. She was nearly parallel-sided and square-ended, and the bottom was cut away into a tunnel along the centreline, the hull being in effect a catamaran; though she differed from any catamarans yet known in yachting – those produced by Nathaniel Herreshoff.

acceptable, but yesterday's and today's are not the same thing. The small raters generally carried a paid hand – or two. In the Solent they would probably have come from Itchen village. An owner might race alone with his hand. It is sometimes regarded as a matter for congratulations that yachtsmen no longer need professionals to help them sail their craft. But the link between the owners and their paid hands, between amateur and professional seaman, held even yacht racing in the smallest boats within the framework of the great sea affair. It was more than a mere sport. Racing helmsmen might escape classification with Kipling's flannelled fools at the wicket and muddied oafs at the goal. They were

Lady helmsmen and little ships.

participants in the small boat seaman's way of life.

A number of lady helmsmen and owners became well known in the smaller classes. In yachting caps only slightly feminized and well buttoned up to the neck they showed again what had already been demonstrated among the males, that good hands for a horse usually brought with them a good touch on a helm. The heavier deckwork undertaken by their great-grand-daughters, together with the language sometimes accompanying it, was beyond them, owing as much to

Above

The 6-Metre fleet racing in 1910. It will be seen that the rig is the conventional gaff rig with topsail.

Left

A Bug, one of a short-lived class but historically memorable as the forerunner of the Star class.

convention as any inherent inability in either sphere. But toughness and femininity were still able to exist side by side. Out on the water, as ashore on the lawns, the sex that was so candid among Society's insiders gave colour to summer days; and a little few amongst those with advanced inclinations found their names associated, in carefully undenied whispers, with royalty; for the Prince of Wales, shortly to become King Edward VII, stood at the social apex of the Solent. Decades afterwards elderly ladies with faces still fresh after a lifelong abstinence from cosmetics, would turn towards the familiar sea scene from club-house balconies full now of strangers, who had never heard of the 'raters' and the races of long ago, yet be aware that distinction was still theirs, and smile at . . . perhaps human gullibility.

When, as told above, the length and sail area rule was discarded, and the first R.Y.A. rule adopted, and later the first International rule in 1906, these rules were used for the small classes. Under the former there was for a time a class of 24-footers, of about the size of today's Dragons. Under the latter rule there appeared two classes that fell readily into the category of small craft, rating at 5- and 6-Metres, and two others that by the standard of the time, though not today, were considered small, the 7- and 8-Metres.

Between 1907 and 1914 there were at least 123 boats built to the four above smallest of the International Yacht Racing Union's classes. This may not seem an impressive number by the standards of today's mass dinghy classes; at the time it represented seven seasons of splendid class racing in thoroughbred yachts, and a wealth of amateur sport of a kind yet new to sailing.

The 5-Metre class never came to much; only five boats were built in the period. The 7-Metre class produced twenty-three boats during the seven years. At that time, early in the evolution of yachts under the new rule, boat sizes in each class was smaller than they later became; an 8-Metre of the pre-1914 era was about the size that 6-Metres later became, the earlier craft, with their gaff rigs and small headsails and spinnakers – the genoa jib being yet unused – having more of their rating in sail area and less in the hull. The 7-Metre was discontinued after World War I.

The 6-Metre class – perhaps the greatest and most superbly performing, and ultimately the most sophisticated small inshore racing yacht ever to have appeared – trails rich memories through the story of yacht racing for forty years; and the boats are remembered today by those who raced them with all the regrets felt for lovely lost things. The rule was modified several times during this period, but there remained throughout the basic framework within which aspiring design technique climbed towards ever greater perfection. The table shows the basic dimensions of a 6-Metre in 1907 and one of the last of them, designed after World War II.

	1907	1947
Length overall	30 ft	38 ft 6 in
Length waterline	19 ft 8 in	23 ft 6 in
Beam	6 ft 10 in	6 ft
Draught	4 ft 3 in	4 ft 1 in
Displacement tonnage	3 tons	4.1 tons
Sail area	507 sq ft	450 sq ft

Between 1907 and 1914, thirty-seven 6-Metres were built. But the evolving form of future yacht racing may be seen in the fact that in this period the biggest of the smaller classes – the 8-Metres – raised no less than fifty-eight boats, or almost as many as the 6- and 7-Metre classes combined, while in the years between 1920 and 1939 the number of 6-Metres built rose to 102 boats, while the number of 8-Metres fell to fifty-two.

With the appearance of these smaller metre classes we find under us the stream of modern yachting bearing us towards the transformed world of the second half of the twentieth century.

Opposite

Catboats sailing on the Delaware, 1874.
Painting by Thomas Eakins.

During the thirty-seven years that William Gardner practised yacht architecture,

25

U.S.

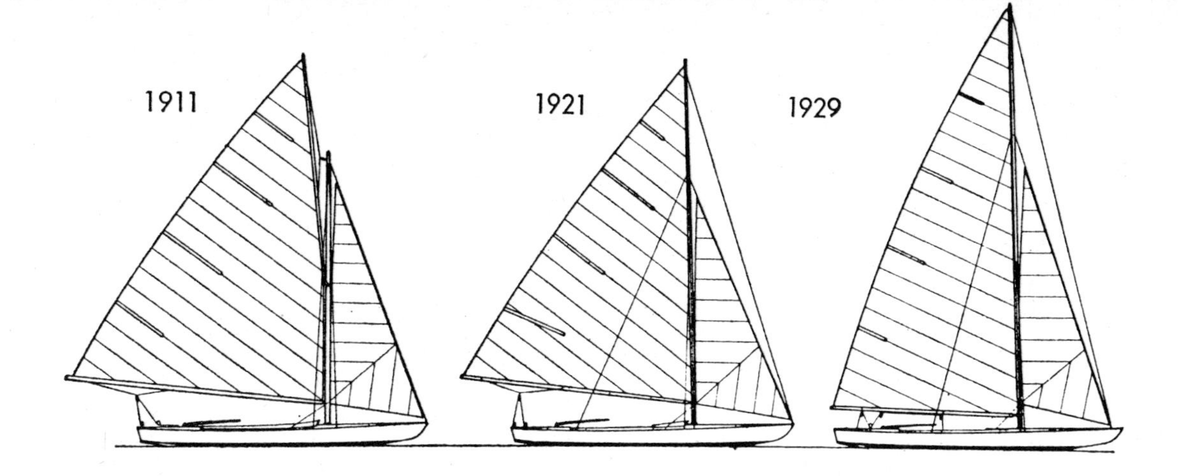

Opposite

Unidentified painting of catboats, in the Peabody Museum, Salem.

he operated from his office at No. 1 Broadway looking over the upper bay of New York harbour. A minor job being performed in the drawing office in 1907 was the production of lines and sail plan for a 17 ft overall keel sloop. This received the unbeguiling name of *Bug*. A small one-design class of Bugs was established on Long Island Sound without stirring particular enthusiasm; but three years later in the Broadway office drawings were proceeding on another similar, though very slightly larger, keelboat. Such an apparently minor piece of design work caused little attention at the time in the drawing office busy with the proper yachts, steam and sail, of the day, except possibly to Francis Sweisguth, the draughtsman working on the plans. Even he could not have guessed that the boat taking shape on the paper under his battens and curves was to become the prototype of the first, and for some time the only, successful application of the one-design principle on a world-wide scale.

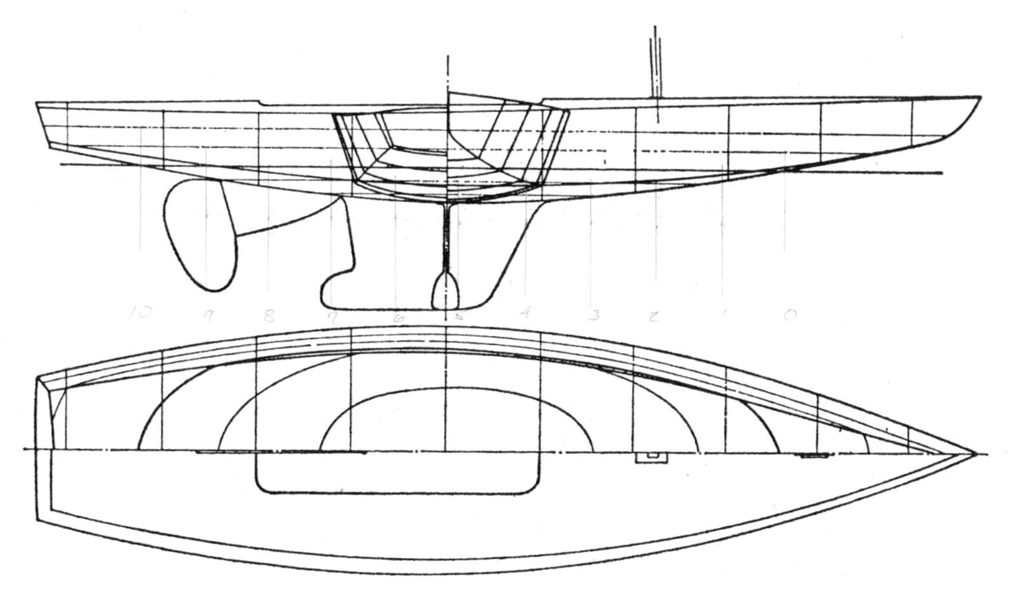

Above and left

The Star class, which came out in 1911, the first successful application of the one-design principle on a world-wide scale, and still one of the great racing classes. The rig above a hull of the utmost simplicity has been changed three times, as illustrated. The Bermudian rig now carried has an old-fashioned air today but suits the hull.

The Bug had become a Star, and half a century later more than 3,500 boats, split into more than two hundred fleets scattered round the coasts of the world, were sailing under the emblem of the red five pointed star on their mainsails. By that time the day of the small boat classes, each registering a dizzying multitude of boats, had arrived. The Stars had not only been the first but was still the most widespread, best known internationally. In the later 1960s at the entrance to the Coliseum, during the New York Boat Show, there stood one of the first Stars to parade itself in the glass reinforced plastics style of the latter part of the century, whose early years had first seen Stars in the honest timber of which still most of them were constructed. Stars were one of the Olympic Games yachting classes until 1968. There were even people who were getting tired of seeing Stars by this time. But in 1911 the future was written in the Stars.

The Star class was designed as an adults' racing boat and grew to strength initially through the drive of George A. Corry, Commodore of the Manhasset Bay Yacht Club, of whom it was later said that the recognition of ability, regardless of financial or social status, was George Corry's contribution to yachting. Trite enough today, and despite the success at that time of the small classes in the Solent, such an idea then had the revolutionary touch. Stars served as the poor man's racing yacht; but as with the small raters in Britain and other craft of high performance, they appealed to and held the loyalty of those to whom the racing of small craft was a fine art in its own right. This again, though self-evident later, was not obvious when the Stars were young.

The boats that emerged on the drawing board at No. 1 Broadway was of the utmost simplicity, a chine hull with almost constant flare along the topsides, a very slight transverse curvature in the bottom under the chine, and an amount of decking that provided a small opening that was in effect a foot-well. When the boat was at rest there was a long though abruptly snubbed overhang forward, which quickly immersed when sailing, and a short counter. The dimensions were 22 ft 7 in overall, 15 ft 6 in on the waterline and 5 ft 8 in beam. The Bugs had failed because at the time they were considered too small for adult sailing – an attitude with a trace of period in it, it may be felt, half a century later. The dimensions of the Star proved widely acceptable as soon as the class appeared. A plate fin carried a bulb of lead at a draught of 3 ft 6 in and the rudder, in a trunk, was hung on a skeg. It was a hull form of the most elementary simpleness, and basically it has remained so.

The rig has been changed three times. The original sail plan was the conventional lug of the day, with the mast well forward, a small jib and the mainboom extending beyond the counter. In 1921 the Bermudian rig was adopted; the proportions of the plan remained the same while the long yard was eliminated. In 1929 the change was made to the present rig, having a taller mainsail with a shorter foot and a jib with a little more hoist. The rig established then remains today, becoming during the period increasingly unfashionable with its big area, low height, small jib, and the lack of either genoa or spinnaker.

The pioneer – Richard McMullen, single-handed in *Leo*, running up Channel in 1857.

8

Riddles of the sands – or mainly cruising

While in Britain the popular idea of yachting was expressed by the Solent, though the fine art of yacht racing was not concentrated only there, one breed of yachtsman, little publicized, was developing something like an antipathy to the Solent. This was not geographical but a matter of temperament, and was discernible all round the coasts, where there appeared many little imitative Solents as the new yacht clubs spread their influence, encouraged racing, found strong support. Meanwhile other people were discovering the pleasure of going to sea under sail in their own unorganized way.

The first man who went to sea on his own, cruising well offshore in small boats, who has left any adequate record of his activities was Richard Turrell McMullen. He started sailing in this way as early as 1850. He died of a heart attack while alone at the helm of his yacht *Perseus*, in mid-Channel on a night of light breeze lit by a clear young moon, in June 1891. The journal *The Field* published an obituary in which it was said (the italics are our own):

'Mr McMullen was unlike any other yachtsman we ever met: we have known men just as fond of the sea as he was, but never anyone who regarded it with such reverential interest. *Yachting and yacht racing in the ordinary sense of the terms had no charms for him.*'

When this was written McMullen had been demonstrating, initially as a pioneer, the art of the amateur seaman for forty-one years. Now, a century and a quarter after he began sailing, it is hard to appreciate how original were his demonstrations and how lonely at first the course he followed, led by his own light. It was a widely held belief, not least among those with some knowledge of seagoing, that landsmen were unfitted to handle small craft far offshore and alone. This was work for men whose livelihood it was, brought up from childhood to the harsh experiences of fishing, coasting, operating the pilot cutters, the ferries and the beach boats that operated between ship and shore. It was not unreasonably maintained that landsmen of relatively sedentary habits not only lacked the rudimentary strength and the toughness inbred through generations in the professional small boat seamen, but that the nature of the skill needed to handle such craft in open water could only come from daily experience begun at an early age and pursued summer and winter. In the lack at that time of evidence to the contrary such arguments were not easily countered by those with enough knowledge of the sea to respect it.

McMullen's first boat was *Leo*, of 20 ft length overall and 3 tons; the next two were larger, the *Sirius* and the *Orion*, respectively 32 ft and 42 ft overall, none of them conforming to their period's idea of yachts. He confined his cruising to the waters of the British Isles, especially the Channel, including a voyage round Britain and one to the Scilly Islands. The man had a genius for the sea enabling him to discern the great truths of seamanship, so hardly learned by landsmen, such as that greatest one of all – that in severe weather it is the shore and not the deeper water that holds danger for a well-found small craft. He made himself from the beginning discover by practice and experience: 'In this manner, getting into scrapes and out of them, I learned more of practical sailing in a few months than I should have learnt in several years if I had a hired man to take the lead in everything.'

Like other fine natural seamen, he was a gifted writer, the first of many to come from the ranks of the cruising men who followed the way along which he led. He wrote thus of the more conventional kind of yachtsmen he knew: 'And when you see a yachting gilt-bespangled dandy, trust rather to your *Seaman's Manual and Vocabulary of Sea-terms*, and do not disgust the gentleman with awkward questions before company. . . .' This comes from *Down Channel*, the first of many books on small yacht voyaging and a classic amongst what is now a multitude. The word yachting, liked or disliked, was spreading its wings to cover new worlds where there was much of the future incubating.

While McMullen was sailing, other men, barely known to the story of yachting, were sailing across the Atlantic in small craft, particularly dories. The readiness to cast oneself on that ocean, even sailing the easier way from west to east, was not yet an activity for other than clear eccentrics. But there was emerging, thanks to these activities hardly even on the outer fringe of yachting as recognized, an appreciation of the fact that ultimate safety at sea did not depend on the size of the craft. While McMullen was showing that the coast and not the sea was the danger, others were revealing the yet unrecognized fact that it was not the size of the vessel that assisted survival when in the oceans. Very small craft could be safe craft.

Captain Joshua Slocum, photographed in 1907.

Captain Joshua Slocum's voyage round the world during 1895–8 has maintained his fame, and that of his boat *Spray*, in the forefront of ocean voyaging annals ever since. He was a Canadian, born in Nova Scotia, who became a naturalized citizen of the U.S.A.; it was the stars and stripes that *Spray* carried round the world. A professional seaman and a shipmaster of twenty years' experience when he set off on his circumnavigation, his son said of him many years later, 'to battle with the ocean was his idea of fun'. But having spent a working lifetime in sail, Slocum could not immediately encourage by his achievement the idea that the average yachtsman and amateur in seamanship might emulate such a performance. Sailing from Boston, Massachusetts, at the end of April 1895, Slocum crossed the Atlantic, via the Azores, to Gibraltar. Thence he crossed the south Atlantic to Brazil, sailed down the South American coast and entered the Pacific through the Strait of Magellan. He put in at the Juan Fernandez islands, and afterwards sailed for seventy-two days westward over the Pacific, eventually landing on a Samoan island. He spent some time in Australia, then sailed north-about and across the Indian Ocean, with two island stops, to Cape Town. He crossed the Atlantic for the third time, via St Helena and Ascension, to Brazil, and afterwards worked up the north coast of South America, reaching Rhode Island three years and two months after he had sailed.

It is to be noticed that he entered the Pacific via the Strait of Magellan. There was as yet no Panama Canal and the rounding of the Horn in a craft of *Spray*'s size, 36 ft 9 in in length overall, and singlehanded, could at that time only have

Slocum's *Spray*, a boat with remarkable self-steering properties, in which the world was first circumnavigated singlehanded during 1895–8.

been considered a suicidal venture. But as readers of Slocum's famous book *Sailing Alone Round The World* will know, the passage of Magellan became a dangerous and at times almost riotous adventure. When the world was circumnavigated singlehanded for the second time, by Harry Pidgeon in the 34 ft yawl *Islander* in 1925, the new canal was used.

Slocum's voyage, like others that were to follow, but unlike some of our own day, was not undertaken with any pre-planned course or time of voyage. He altered his intended course twice, by not entering the Mediterranean and not sailing to the south of Australia. It was enough to achieve a circumnavigation alone, to treat it as a small yacht cruise, though of an unprecedentedly long and dangerous length. Again, like later voyages but not those of the present time, *Spray* was not equipped with any form of self-steering gear. But her unusual broad and shallow hull under its yawl rig proved to have remarkable self-steering qualities when rightly trimmed. This, at the time, was considered to be a remarkable feature of the yacht and the voyage. When *Spray* was fitted out for her last voyage at Herreshoff's works in Bristol, Rhode Island, it is recorded that Nathaniel Herreshoff, then at the height of his fame, spent much time examining the yacht and questioning her skipper.

On the voyage that followed *Spray* was lost at sea in unknown circumstances, the general supposition being that she was run down. So the pioneer of singlehanded ocean sailing may have been drowned under the circumstances that still today cause many people to regard singlehanded sailing as fundamentally unseamanlike, owing to the impossibility of maintaining a proper lookout.

During the course of World War II one of the fleet of American Liberty ships was named Joshua Slocum.

Captain John C. Voss was sailing his sloop *Xora* in 1898, the year when Slocum returned from his circumnavigation. A Canadian, he was like Slocum a professional seaman of wide experience under sail when he turned to small craft. And like the amateur McMullen he brought to their handling an enthusiasm for experience and experiment that in the broad sense contributed to man's philosophical knowledge of the sea, and in the narrower sense helped to open the wide oceans to small craft voyaging.

His experience and keen mind makes all the more odd his choice of craft for the voyage that made him famous. The hope of financial profit encouraged his selection of boat. A journalist had suggested to him that £1,000 might be made if a boat smaller than Slocum's *Spray* were sailed across the three oceans, Pacific, Indian, and Atlantic. Voss determined to make the voyage yet more remarkable by selecting for the purpose a dugout Indian canoe. Stunt and the profit motive may have vitiated the pure Corinthian spirit of conventional yachting, but this was to become usual for such voyagers, developed to a highly organized system of sponsorship by newspapers and commercial firms by the 1970s. The fact that such ocean exploits can rarely be made by those fit and competent enough to do so without such financial aid is justification enough.

Voss's *Tilikum* was a fifty-year-old canoe 38 ft long overall, 30 ft on the keel, dug from a log of red cedar. A product of one of the most ancient boatbuilding techniques, Voss contrived to make modifications that converted the hull into one suitable – at least in his competent hands – for so exacting a task. The inherent fragility of the dugout he palliated by fitting transverse oak frames internally. The dugout's inherent lack of sail-carrying power, owing to its narrowness – *Tilikum* had a beam of only 5 ft 6 in – was corrected so far as possible by fitting inside three tons of lead ballast; but such a boat could not carry any conventional rig. Sail was set on three low masts in a gaff foresail and main and a triangular, or leg of mutton mizzen, with a staysail set forward. The smallness of these four sails, just enough to propel so easily driven a hull, was one reason for the success of the curious little ship. Another was her light draught, of no more than two feet. Of comfort on board there was inevitably little. Voss decked the hull and fitted a cabin top covering a cabin 8 ft in length. But comfort in a small craft in the midst of the ocean is judged by other standards than accepted ashore. How many people have found their elaborate cabins becoming a nightmare shambles once at sea?

As late as 1880 there was no organization concerned with the interests of cruising yachtsmen. Racing inevitably demands organization, rising from the club to more comprehensive bodies, which by this date reached to the Yacht Racing Association in Britain. Cruising people were meanwhile growing in numbers and going their own way. This was the essence of the activity. In the soul of every dedicated cruising man was engraved a feeling expressed by a later Poet Laureate about the lonely sea and the sky. They were a contemplative people, perhaps with a tendency to introspection: and it was even suggested by bigots that they had higher intelligence than was usually found among the contending racing helmsmen. Certainly there was a schism between the worlds of cruising and racing, which the later twentieth century was to iron out almost completely; though even in the 1950s there were a few who still felt inclined – though they may have fallen short of the action – to follow the example of one of their number, who exhibited below the name of his boat the initials M.O.B.Y.C., standing for 'my own bloody yacht club'.

The Cruising Club was founded in the autumn of 1880 at the Lincoln's Inn chamber of Arthur Underhill, later Sir Arthur and Commodore of the club for fifty years (1886–1936). The first Commodore, the Revd A. van Straubenzee, a retired naval officer, then owned a 10-ton cutter; Underhill a 2½-ton sloop; and the boats of the club's small membership were likewise modest. By the following year there were fifty-one members. In 1882 arrangements were made for the printing of a journal for private circulation containing accounts of cruises by members; this became the Cruising Club Journal, an anthology of cruising logs that remains an annual production today. The qualities sought in these logs, as outlined by Sir Arthur Underhill, express clearly the objects of the club. A cruise must show good seamanship, navigation and pilotage, and be undertaken in a well equipped yacht. Credit was given to long cruises provided they were within the carefully assessed capability of the yacht (at that time the oceans were considered to be beyond the capacity of the best small craft) and members were encouraged to explore the lesser known coasts and harbours, giving pilotage information on them.

In pursuit of these objects, so different from those of the racing fleets, a series of cups were offered; first came the Club Challenge Cup, presented in 1896. In 1955 this was won for a voyage round the world in a yacht of 8 tons – Mr and Mrs Hiscock's *Wanderer III* and that such a voyage could have been undertaken with something like insouciance owed not a little to the lessons that the Cruising Club had been instilling for three-quarters of a century. Next came the Romola Cup and the Claymore Cup, presented in 1909 and 1911, followed by the Founders Cup. The club received the Royal Warrant in 1902 and appeared under the name by which it has become so widely known – the Royal Cruising Club. Twenty years had still to pass before a counterpart to this club was to be founded in the U.S.A.

Five years after the Cruising Club, the Cruising Association was formed in

Britain. This was a markedly different kind of body from the R.C.C.; an association, not a club, of unlimited membership with the object of providing facilities for all who cruise; these developed in various directions and included the issue of a handbook of sailing directions for the ports of Britain, including Ireland, and north and west France; the publication of harbour plans; the production of a list of recommended boatmen at various ports; the collection of a library in a London clubhouse; the presentation of annual challenge cups, the first of which was given in 1911.

Cruising in relatively small yachts was becoming, like yacht racing, a fine art, though one of a quite different *genre*. Claud Worth's book *Yacht Cruising*, published in 1910 and still in print, provided an anatomy of the art. In the preface to the first edition the author wrote: 'Yacht Cruising may be either a pastime or a sport. To sail from port to port by easy stages in picked weather is a most pleasant *pastime*. To make an open sea cruise in a seaworthy little yacht, neither courting unnecessary risks nor being unduly anxious as to weather, and having confidence in one's knowledge and skill to overcome such difficulties as may arise, is, to one who loves the sea, the most perfectly satisfying of all forms of *sport*.'

In his book Claud Worth laid bare the many skills, crafts and practices, the many branches of expertise that made the all-round seaman and handler of craft in the broadest sense. When *Yacht Cruising* appeared it had no counterpart. There was little published about handling small craft in bad weather. The work of rigging, fitting out, maintenance, the commonplace operations of caulking

Direct descendant of the first European yachts, and rich in ancestral characteristics, the Dutch boier continues to be a type of cruising yacht favoured by some.

and painting, the judging of good timber for masts, spars and planking – there was little to guide the amateur towards such skills, which had then to be learned largely from professional craftsmen, who tended to be inarticulate men of action. The many skills, including navigation and pilotage, formed a big corpus of knowledge and experience which Claud Worth exposed, and since then many have expanded, to produce the educated amateur seamen of two wars and today.

In 1903 Erskine Childers' *Riddle of the Sands* was published, and the young art of yacht cruising was given literary expression. It was a novel with a didactic purpose, but the problem of national defence with which it was concerned soon became a period piece. Not, however, the book. In 1961 the literary staff of the London *Sunday Times* made a selection of ninety-nine great prose works that have appeared during the twentieth century. Afterwards readers were invited to bring the total up to 101 by suggesting two other worthy books, one of fiction and one non-fiction.

It was interesting to find that *The Riddle* was high on the suggested fiction list; out of 284 works of fiction suggested by readers, *The Riddle* came fourth in the number of votes it received. A conclusion to be drawn is that in 1961 a high proportion of the readers of a national newspaper were more than casually acquainted with the art of yacht cruising, which would certainly not have been the case in 1903. But even in the latter year, amongst the general public who read the book for its story and its propaganda, there were a few who recognized in it a fine literary expression of the quiet but not unadventurous activity of 'messing about in boats' – adventure, which Childers defined as 'the gay pursuit of a perilous quest'. The hero of the book is *Dulcibella*, the tiny centreboard cruiser converted from a ship's lifeboat.

Cruising in relatively small yachts had at this time a particularly strong appeal to men of the more learned professions, with the law prominently represented amongst them. The popular novelist, the late Nevil Shute Norway, in his autobiography *Slide Rule* writes of the headmaster of his preparatory school in Oxford during the early years of the present century, a school with an unusually high scholastic record. The founder and headmaster, C. C. Lynam, as described by Shute, seems to typify one dominant kind of cruising yachtsmen, men of intellectual distinction and not too conforming a cast of mind:

'The headmaster was known to everybody as The Skipper because yacht cruising was his passion; he was a big, red-faced, laughing man with white hair that was seldom cut and curled about his ears. . . . He had a succession of three or four sailing yachts that he called the *Blue Dragon*, in which he used to cruise around the north of Scotland, the Hebrides, the Orkneys and the Shetlands, during the Easter and summer holidays. Towards the end of my time at the school he took a term off and sailed his boat across to Norway and up the coast to the North Cape.'

Lynam was rugged in his choice of cruising areas up in the Viking seas; but if more ambitious than many he was also typical of other men of culture who cast off the formal shoregoing clothes of the period and reappeared in the blue jumper of the fisherman at the tiller of a small craft.

There was often a wide chasm of taste between the cruising and racing yachtsman, nowhere more apparently unbridgable than between the attitudes on the east – Essex and Suffolk – coast and those of the south, where the tone was that of the Solent. The conflict of taste found an outward expression in sartorial matters, for clothes make the yachtsman as well as other men. Even the term 'yachtsman' itself had a touch of pretension in east coast ears. Were not the men

Left

In a cruiser at Le Havre in 1913 the paid hands face the camera. A professional crew of four in a yacht of this size, then considered small, was not unusual.

Opposite

An impression of cruising under conditions rich but rough by R. C. Woodville.

who were at home in the lumpy shoalwaters, where the ebb made little draining gullies of the fairways, raised tidal islands and left shining fields of mud, practising the small boat seamen's old trade? They were 'amateur seamen' rather than 'yachtsmen'. The landlord behind the counter of a quayside bar parlour might be preferable to the yacht club's steward, and real salty talk to class racing gossip. The familiar yachting cap of the Solent, especially if topped by a starched white cover, was something to raise an eyebrows in a period when small boat cruising contained a happily bohemian element. A full yachting rig-out of reefer jacket and white flannels, or blue trousers matching the jacket and waistcoat, the precise black tie and pin, were the liveries of the stage yachtsmen, not competent amateur seamen.

There was prejudice and a little that was unamiable in such ideas, and a touch also, perhaps, of envy; there was also the expression of two temperaments. Professor Sir Walter Raleigh once wrote what might have been a parody of the dedicated cruising yachtsman's attitude to the Solent, though he probably did not intend this:

'The so-called yachtsmen of Cowes, it seems, have champagne lunches in the Solent, and then spread canvas and take a two hours' run when the breeze is fair and light. They meet the same people at Cowes as they meet later at the grouse, and this life which they make to resemble Heaven must be a live facsimile of Hell. Different clothes and the same well-fed, carefully exercised bodies, the same bored minds tired of wondering whether passion will ever come their way.'

None of this was for the cruising men, who were sure that he who 'goes to the sea in a large boat, run by other men and full of comforts, can only do so being rich, and his cruise will be the dull round of the rich man. But if he goes in a small boat, dependent upon his own energy and skill, never achieving anything with that energy and skill save the perpetual repetition of calm and storm, danger undesired and somehow overcome, then he will be a poor man, and his voyage

will be the parallel of the life of a poor man – discomfort, dread, strong strain, a life all moving.'

Thus wrote Hilaire Belloc when he was at the height of his fame as a trenchant, controversial author. He was also a sailing man dyed deep in the blue of cruising, which he expressed with an energetic prejudice that earned his book *The Cruise of the Nona* an almost inevitable place on the cruiser's cabin bookshelf alongside *The Riddle of the Sands*. Belloc was convinced that yacht racing led to the devil. Nobody bemused by the magic of the swatchways, held slave to the solitary creeks away from it all, could have raised such an almost pathological loathing as he did for what Joseph Conrad had called 'the fine art' of yacht racing: 'For no one can doubt that the practice of sailing, which renews in us all the past of our blood, has been abominably corrupted by racing . . . Remember, and write it down: *"Cruising is not racing."* If your boat is a home and a companion, and at the same time a genius that takes you from place to place . . . a good angel, revealing unexpected things, and a comforter and an introducer to the Infinite Verities – and my boat is all these things – then you must put away from yourself altogether the idea of racing. . . .'

And nobody could express the cruising man's *credo* more emphatically and poetically, if slightly absurdly, than this. A founder member of the Cruising Club of America, Frank P. Draper, was to express the same attitude with less frenzy when, in 1923, he raised his objections to the Club's support of ocean racing, for which it subsequently gained its international reputation:

'I am of the opinion that the whole spirit of racing is radically opposed to the

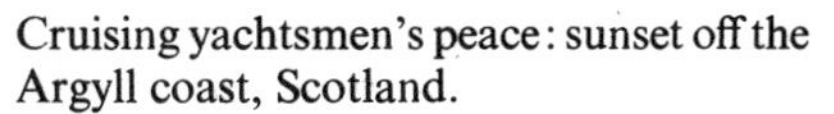

Cruising yachtsmen's peace: sunset off the Argyll coast, Scotland.

spirit of cruising . . . The real fact must always remain that the cruising boat cannot be a racer. It is a contradiction in terms.'

Frank Draper was wrong in regard to the boats concerned. With regard to the spirit of the two activities, his perception was sounder.

The Firth of Clyde is the largest and most splendid area of water offered by the British coast for yacht cruising. Clyde yachtsmen have always suffered, perhaps, from a mild schizophrenia, uncertain whether their home waters are a stronghold of racing or cruising, and themselves doubtful whether to answer the starting-guns or enjoy the ever-changing Firth, where deep water with few shoals and little tide runs between islands and curls among landscapes that may be softly pastoral or dominant with vaulting, heather-clad hills – the country for which exiles have wept. In later years, when after two world wars, ocean and offshore racing became ascendant in yacht racing, it made little more than a dent in the cosmology of the Clyde. Southerners go offshore racing, it was then said, because they have no lovely, uncrowded local waters to tempt them to stay at home.

By a river so busy with shipbuilding and maritime affairs, there had always been an indigenous population ready to take to the water, and a higher proportion of it may have done so than of the Solent people. But the Firth, on the edge of the Scottish Highlands instead of the English Home Counties, missed the fashionable attention given to southern yachting, and the considerable Clyde yachting activity has less history to record, though it may have passed days more fully satisfying. Certainly this was so for those who cruised. It is, however, typical of the general situation in yachting that while the first Clyde Week was organized in 1877 and there were numerous and active yacht clubs, there was still in 1909 no organization for the purely cruising man.

In this year the Clyde Cruising Club was formed with the objects described, perhaps inadequately, as 'to encourage cruising and foster the social side of sailing'. By 1910 was a membership of 113 owning forty-nine yachts with a tonnage of 520 Thames measurement, which by 1914 had increased to 194 members, and eighty-eight yachts with the tonnage of 1,206. The average size of yacht, of about 15 tons T. M., with several far above this average, indicates the style of Clyde cruising. While the yachts were small compared with so many found on the Solent, they were, even in terms of the day, craft of some size and comfort.

Norway, a nation of seafarers, came relatively late to yachting. Appropriately, when she had done so, it was yacht cruising and a certain type of eminently seaworthy yacht that brought Norway international recognition. Later, it is believed that Norway had a higher proportion of yachting and sailing people in her population than any other nation.

Colin Archer of Tolderodden designed and built what is believed to be the first Norwegian yacht in 1873, and it was Archer's reputation that initially brought Norwegian yachting to the attention of Europe. Had he not been born by the sea, and on a coastline studded with harbours, surrounded by a seafaring population which could look outwards only towards the dangerous waters between the Naze and the Scaw, in the seaward approaches to Oslo with its fjord and the inner leads, Colin Archer might never have become absorbed in the world of small craft. But Larvik lies at almost the extreme point of the deep vee formed where the southern end of the Scandinavian peninsular is cleft and the Skagerrak runs up to what was then Kristania Fjord; a coastline inclining north-west, peppered with old towns and villages and islands, where activity was largely seafaring. The five miles of Larvik Fjord run inland approximately north-west from this coast. Some sixty miles down the coast is Arendal, where the Archers had first landed when the family emigrated from Scotland. Twenty miles in the other direction, among strewn islands with villages upon them, is Tönsberg, and across the wide opening of the fjord, scattered with more islands, is Fredrikstad, with Hankø near by, today regarded as the Cowes of Norway. But there were no yachts in Archer's youth, and several decades were to pass before he designed and built the first Norwegian yacht. From the entrance to Larvik fjord, Oslo fjord drives about eighty tideless miles inland. On the eastern side of Larvik fjord, by the spired church in the town, with its grounds running down to the water, where Archer's building slips were later to be laid, stood Tolderodden, on a rocky promontory to the south-east of the small town.

Colin Archer devoted himself primarily to improving the design and construction of the pilot boats, fishing and life-saving craft of his country. At his funeral Johan Anker, when laying a wreath from the Royal Norwegian Yacht Club on the grave said: 'As long as there are sailing craft on our coasts so long will his memory live. For all of us younger men he was an ideal – this fine old man.'

It was against the background of this nationally vital activity of improving the breed of Norwegian working craft that Archer's concern with yachting must be considered, and the types of cruising yacht associated with his name were the direct outcome of his experience with professional seamen's boats. Ten years after producing Norway's first yacht he became a founder member of the (Royal) Norwegian Yacht Club, of whose management committee he was a member until his death in 1921. There followed a number of yachts designed and built by him for owners in Norway and Sweden, and these finding their way round the European coast gained the reputation of what became broadly known as the 'Colin Archer type'. While a great number of the type came from Archer's drawing board and his yard at Tolderodden, the term came to be applied to many other yachts of Scandinavian type with big beam, heavy displacement and the characteristic double-ended Scandinavian form, having a pointed stern and rudder hung outboard. Archer's interest in the theories of naval architecture and his publications on these subjects in *The Field* in London and the Institution of Naval Architects, brought him recognition as a yacht designer unique for one who had never designed a racing yacht and who had concentrated on that then unfashionable area of yacht design, the design of the smaller type of cruising and deep-water craft.

There was a strong body of opinion in cruising circles that the Archer yachts were the finest yet produced for their purpose, and it was still held by some experienced seamen in the years after World War II. During 1954–6 Group Captain and Mrs T. H. Carr with a crew of up to six at certain times sailed widely in the north and south Atlantic, visiting the West Indies, Bermuda, Panama and South Africa. The yacht was *Hawfruen III*, a large Colin Archer yacht of 71 tons Thames measurement built in 1897. Some yachts were conversions of Archer's working boats. One was Erling Tambs's *Teddy*, a pilot boat built in 1890, in which he and his wife sailed from Oslo to Auckland, New Zealand, via the Panama canal in 1928, a story told in Tambs's *The Cruise of the Teddy*. Another was Ralph Stock's *Oeger*. She had been built in 1908 as a lifeboat for the North Sea fishing fleet, and later converted into a yacht. Stock found her in Devon during the 1914–18 war and soon after the war set out on a voyage with his sister and one or occasionally two others, from Brixham to the Pacific Tonga Islands via the Panama canal, putting in at Vigo, Canary Islands, the Barbados, Galapagos Islands, the Marquesas, Tahiti, the Savage Islands and Tonga. It was in an important respect a pioneer yacht voyage; the first considerable ocean cruise in a yacht able to carry in modest comfort a small crew and make the sea its home, together with the scattered ports along half the earth's circumference. It was something different from the epics of the singlehanders, and suited to less eccentric tastes. Ralph Stock described his voyage in his book *The Dream Ship*; and yachtsmen have been following in the *Oeger*'s wake – a small, select company of seamen – in gradually increasing numbers, since 1920.

One of Archer's last letters was written to Stock after he had arrived in the Tongas. He was in his eighty-ninth year. Nothing was more fitting in his life than that he, who had been the first highly qualified marine architect to devote himself exclusively to the design of cruising yachts, generally of small size, should also have produced a boat for Erskine Childers, the first person to give literary expression to the art of yacht cruising. In 1905 he had designed and built the *Asgard* as a successor to *Dulcibella*.

Rescue boat, *Svolvaer*, 1897, designed by Colin Archer.

9 Steam and internal combustion

Steam yachts entered history under a cloud of disapproval as thick as their funnel smoke. Their day was not long, the period of full bloom extending for little more than half a century. They had – and now their memory retains – the appeal of any extreme manifestation. Among their number were some of the loveliest expressions of marine architecture and some of the craziest of marine follies. They were sometimes hardly less than noble in their beauty both inside and out; others merely showed the horrible results of unbridled extravagance and an idiot guiding mind. Good and bad were expressions of a kind of society now seeming more remote than it actually is, and which, being dead, has assumed a mellow glamour.

In this respect it is like other social aspects of nineteenth-century yachting, though raised to a more vivid plane. But when the American yacht architect L. Francis Herreshoff, son of the hard-headed 'Wizard of Bristol', described in 1958 the days of the British steam yachts' apogee thus, we may feel the power of lost yesterdays to touch imaginative minds and become a little distorted:

'The largest and handsomest of these yachts generally had an afterguard composed of the most aristocratic ladies and gentlemen in the world and often included royalty. As a rule these fine vessels were owned by landed proprietors who in the old days had secure positions in society. Many of them were lords, dukes, and even princes, owning large tracts of land which in some cases had been granted to their ancestors sometime between the Conquest and Henry VIII. The ladies on the English steam yachts were often pure Anglo–Saxon beauties who could trace their ancestry back to the Domesday Book and whose earlier ancestors had probably been Vikings, as their looks and love of the sea indicated.'

This is a picture of steam yachting on the same level of truth as a portrait by Van Dyck. Here is an impression of the same thing, in the manner of a caricature by James Gillray:

'Life on board the larger yachts in the United States was, at times, more lively than on British ones. First, yachts could be moored right in New York City, where facilities for gaiety were present. . . . Secondly, several owners enjoyed their newly acquired wealth in a rather strenuous manner. One yacht was famous (or infamous, depending on the point of view) for the frequent *danse du ventre* performances given on her after deck. Another had a luxurious owner's cabin complete with an electric console control board. Push one button and soft music was heard. Push another and a small bar opened into view. Push a third, and the wall beside the owner's bed rolled away to find the guest of the evening, having retired to the assumed sanctity of her private cabin, revealed in a bed beside that of the owner. Sherman Hoyt once told of an incident while waiting to go ashore from a yacht to church. The Sunday calm of Newport was rent by a scream from an adjacent yacht. This was quickly followed by the appearance on deck of a curvaceous blond in a filmy nightgown, screaming "You won't ——

Under the domed skylight, the cabriole leg chairs, the ormolu hung dresser, the long-case clock, the rich joiner work, set the scene for Victorian meals on board a steam yacht *c.* 1880.

me again, you S.O.B." as she plunged overboard. There was more screaming when she realized she couldn't swim! As the waiting launch made a very public rescue in which the nightgown of the rescued kept slipping off the body of the rescued, a huge bearded man, completely nude, admonished the launch crew to "Let the double-crossing bitch drown!"'

The above story is told by Erik Hofman in his *The Steam Yachts – An Era of Elegance*. It may be agreed that its heroes and heroines had Viking characteristics no less pronounced, if less engaging, than the charming people described by Herreshoff. Owners of steam yachts had one thing in common amongst their infinite variety – exceptional wealth – and the products of wealth are likely to be more interesting, in one way or another, than those of the tight budget. Among the ownership of large steam yachts there was to be found tycoonery rampant. But there were also men who would have turned anyhow to the sea and yachts for pleasure, and who, being rich, did so in the grand manner.

Initially, however, there was not considered to be much grand about this aspect of yachting. In 1829 the Royal Yacht Squadron, as the custodian of good form in British waters, refused to accept a proposition that the club should receive steam yachts into its fleet. When, fourteen years later, it relented in some degree and decided 'that steamers belonging to the Squadron shall consume their own smoke', this flutter of humour became held as an example of reactionaryism. The Squadron's attitude may not have been as far sighted as a few of its members, but in the climate of the day it was not unreasonable. Yachting was still close enough to its origin to be sensitive about the two objectives that as a sport brought it prestige; to promote seamanship and improve sailing vessels. In 1828 only twenty-four years had passed since Trafalgar. In 1844, when the Royal Yacht Squadron had accepted steamers so long as they exceeded 100 horsepower, the ships of the line, or battleships, in the navies of the world were still three-masted sailing ships essentially similar but larger than those that had fought at Trafalgar.

Steam navigation was still in the heroic age, but it was certainly not conceded, by merchant and naval seamen as well as yachtsmen, that its heroes fitted any acceptable heroic mould. Paddle-wheel steamers with tall chimneys were running ferry services on a few of the busier rivers in the New World and Old. The *Savannah*, belonging to Colonel Stevens of New York, became in 1819 the first vessel fitted with an engine to cross the Atlantic, though the achievement owed little to the machinery; and though in 1825 a ship from Britain reached India under sail helped by steam, it was questioned by few that a steamer unequipped with sail could never be a reliable ocean-going vessel. In 1840 the barque-rigged *Britannia*, the first Cunarder, established the regular steam navigation service across the Atlantic. Nineteen years later there was being launched for the British navy the last wooden wall, a first-rate three-decker, direct descendant of the *Victory* herself, outwardly much like her but concealing within an engine like a vice. But six years before this, in 1853, the Royal Yacht Squadron had withdrawn all rules limiting the use of steam in club yachts. The period of the steam yachts' much harassed probation was over; the period of approbation was on the doorstep.

There are no recorded steam yachts in the U.S.A. until 1853. Beyond the light of history there may have been a few small vessels with steam used as yachts in British waters before the first of which there is record, which was the *Menai*, of 400 tons, belonging to Thomas Assheton-Smith, and which appeared in 1830. Assheton-Smith has already appeared in these pages. In 1829 he decided to end his sailing career, and also his membership of the Royal Yacht Squadron. How much the latter move was owing to an acrimonious dispute he was engaged in with his fellow member Lord Belfast, how much to the attitude of the club towards steam yachts, can no longer be gauged; but in September 1841 the *Sporting Magazine* announced that 'Mr Smith has determined to take his future aquatic excursions in a steam vessel of extraordinary power. . . .' The clouds over steam yachts were thick for more reasons than one at that moment; and the smoke sometimes laid above the Solent by the new steam paddle boat services was the sky sign of the hot feelings amongst the yachtsmen of the area.

Assheton-Smith had built eight steam yachts of various tonnages before the first one on record appeared in the U.S.A.

Steam initially gained respectability in yachting through the great auxiliary sailing yachts which preceded the brief elegant age of the clipper-derived pure steam yacht, while some large sailing-cum-steam yachts persisted into the last decade of the century.

It was in 1856, when the Squadron's rule against the admission of steam yachts was wholly rescinded, some half-dozen members had auxiliaries fitted to their sailing vessels. In a modest way the age of auxiliary mechanical power had opened. By the 1890s there were at least two yachts rigged as three-masted ships; the barque was more usual; and in some vessels the funnel was collapsible or telescopic. Sometimes the canvas was adequate in area for a good performance under sail; but many rigs were deficient in area, as was so in most contemporary naval vessels; performance under sail was poor. At the other end of the scale was such a vessel as the three-masted schooner *Atlantic*, whose race across the Atlantic with a pure racing cutter is described later. *Atlantic* was a sailing ship, her triple expansion steam engine definitely auxiliary. Perhaps in no yacht of the period were sail power and steam so equally balanced as in the famous *Sunbeam* of Lord Brassey, particularly noted for her round-the-world voyage during 1876–7. *Sunbeam*, 170 ft in length overall and displacing 576 tons, was a three-masted topsail schooner, though she could set a considerable array of square sails. Under the 350 indicated horsepower of her compound steam engine with a Scotch boiler she could make 10.3 knots. During her circumnavigation *Sunbeam* covered 15,000 miles under sail out of a total of 27,800 miles. Her best day's run under steam alone was 230 miles; under sail 299 miles. *Sunbeam* was a competitor in the Kaiser's Transatlantic race of 1905, but she and other great yachts like her, with their nearly equal proportions of sail and steam power, could be no match for a sailing auxiliary such as the above *Atlantic*. Some words of Lord

The crew of *Cassandra*, 64 strong, included the chef, assistant chef, nine stewards and seven officers. *Cassandra* was 287 ft in length overall, of 1,227 gross tons.

Brassey written at about this time recall the ideas of an era now seeming so remote: 'Though the fashion of the hour has set strongly towards steam propelled vessels, the beautiful white canvas, and the comparatively easy motion of vessels under sail, will long retain their fascination for all pleasure voyaging. It is pleasant to be free from the thud of engines, the smell of oil, and the horrors of the inevitable coaling.'

By this time most steam yachts had wholly discarded sail, only a few retaining vestigial areas for steadying purposes. But as late as 1898 one of the largest steam yachts built in the U.S.A., the *Aphrodite* of 1,823 tons displacement, came out as a three-masted barque. But like the naval ships, her spread of canvas was inadequate for the size of hull. Her masting and sparring was removed in about 1910. Steam yachts by then flourished their once-loathed smoke with the *panache* of a plume.

But though sail might have become vestigial in the steam yachts, or disappeared altogether, their external grace came directly from what was generally regarded as the most perfect form of ocean-going sailing vessel ever produced – the clipper – which, according to the poet John Masefield was a form of sea beauty man had ceased to build. But in the steam yachts something of that beauty, though bereft of sail, persisted for a while. It was a frankly derivative form growing more and more anachronistic; but it retained its place until the 'thirties of the next century.

Lofty masts were uncrossed by yards but swung high the wireless aerial; clipper bows with some scroll work carried bowsprits for appearance only; long, lazy counter sterns reached out with the elegance of a dandy; and the functionally essential funnel, in the earliest steam yachts an object of shame to be disguised as far as possible, became a proud feature aesthetically integrated into the ships' gracious profiles. The large mechanically propelled yacht emerged into her brief opulent reign in the dress of the sailing clipper and wore it in a maritime world changing radically around it.

But for the discipline imposed by the sea on the rich as well as the rest, such beautiful and efficient ships could never have emerged. On shore the Gothic and Palladian could, and often did, go mad. Quite ordinary Victorian English gentlemen had their houses built in a way that they thought might have suited a Norman baron flushed with the success of the Conquest; or at Newport in Rhode Island bluff American tycoons, as though they were sinuous Italians, sheltered themselves behind the pillars and porticoes of imitation Campania villas. While the steam yacht lacked any well-based tradition of architecture to guide or mislead, the sea was always there to exercise some control over the wilder shores of fancy. So fancy did not go beyond the graceful, the exciting, but also the proved and respectable clipper form which had earned the sanction of the oceans.

A very few yacht architects on both sides of the Atlantic were responsible for the finest steam yachts, and initially it was the British who led the way in both quality of design and construction. British yachts tended to be too slow for American tastes, their interiors too dark and perhaps not ornate enough. This did not prevent Americans ordering yachts from the Clyde designers and yards in big enough numbers to encourage the application of legal means to stop the traffic, culminating in the Payne Bill passed by Congress in 1897. These measures, including the last, were not successful.

It is convenient at this point to complete the story of the steam yacht and so form a link between the last days of the pre-1914 yachting world and those of the inter-war years, 1919–39, during which the last few steam yachts were built.

Amongst those who were attracted towards the ownership of a steam yacht, there were not a few who approached the matter with the attitude of the U.S.A. Steel Trust's President, who at the beginning of the present century asked Clinton Crane to design for him the biggest yacht in the world. After some research Crane determined that such a vessel must be 350 ft in length, though this did not

The third *Savarona*, the largest privately owned yacht with a gross tonnage of 4,646 tons. Owned by Mrs E. R. Cadwalader from 1931 to 1938, she subsequently became, and remains today, a school ship in the Turkish navy.

exceed the lengths of several royal yachts: the German Emperor's *Hohenzollern*. the Russian Emperor's *Standart*, the British Queen's *Victoria and Albert* (third of the name). But the royals apart, the yacht envisaged by Crane, which was ı built, would have exceeded in size any steam yacht built until 1931. As late as this in the history of steam yachting there appeared, briefly under private ownership, the third *Savarona*, designed by the American firm of Gibbs and Cox for an American lady, Mrs Cadwalader, and built by Blohm and Voss. J. P. Morgan's fourth *Corsair*, last, largest, if not quite the loveliest of that brilliant sisterhood, which had come out a year earlier (1930) was otherwise the biggest steam – or indeed motor – yacht to have been built for a private owner; and she was 343½ ft long (2,142 tons gross) which was close to Clinton Crane's specification. But *Savarona* was 408½ ft in length and 4,646 in gross tonnage. She also carried the largest crew of any steam yacht, the royals again excepted, numbering of 107, or about the number of the larger yachts in the richly servanted days pre-1914. The owner used the yacht for two years, and five years later she went where alone so splendid, extravagant, and by this time eccentric a creation could go, into government service, that of Turkey. The President of this small nation then found himself possessed of a yacht rivalling in size that of the King–Emperor of the still surviving British Empire.

The last of the steam yachts in the clipper tradition tended to lose something of the nearly perfect grace found in their earlier sisters. The fourth *Corsair* mentioned above was larger and more spacious in her cabin appointments than the third, which might be claimed to have been the most finely proportioned of all steam yachts. The later yachts were given more freeboard and their long deckhouses were sometimes on two levels. They were better ships than the older but lost some charm. However, in 1930, a vintage year for the type, the Watson-designed steam yacht *Nahlin* was put in commission. There are those who might advance her claim to rival the third *Corsair* in aesthetic power; she

Above

The third *Corsair*, belonging to J. Pierpont Morgan, may be considered by some judges the most perfectly proportioned of all steam yachts in the clipper style. She was built in New York in 1899. Later yachts in the same clipper style had more elaborate deckhouses and accommodation on the same length, at some cost to their harmony of form.

Opposite top

Nahma, of 1,740 tons gross, was a larger yacht than *Corsair* (above), and with her higher topsides was able to provide more accommodation. Designed by G. L. Watson in Glasgow for American owners, *Nahma* lacked some of the grace that this master of steam-yacht design usually achieved. It is to be noticed that both *Nahma*, built in 1897, and *Corsair* carry steadying canvas.

was certainly the larger ship with more accommodation.

By this time experiments in a new style had been proceeding, though they failed to achieve beguiling charm. The steam yachts *Restless* and *Thalassa* by the master George Watson had been commissioned in 1923 and 1924, and they both had straight stems and cruiser sterns, and the sheerline of the latter was flat and graceless compared with the steam clippers.

In 1913 *Pioneer*, the first big diesel-engined yacht came out, designed and built by Camper and Nicholson. By the same firm came *Sister Anne* in 1928. In these two vessels the search for a new idiom of design appropriate to the large diesel yacht was being tentatively sought; they struck the eye of their day but could not satisfy for long. When in 1930 the Watson-designed *Virginia* was commissioned, a diesel yacht but retaining the steam clipper style, unadventurous eyes rested happily on her. But as the decade of the 1930s rolled on to war the new style of large diesel yacht, in the shape of such vessels as *Trenora* and *Shemara* from Thornycroft and *Philante*, Camper and Nicholson, revealed that a new and underivative style had been evolved to suit the new mechanical system.

In the last decade of the old century and the beginning of the twentieth a new branch of yachting was about to emerge. For a brief spell of time the relatively small steam launch became the fastest means of free locomotion known to man – faster than the horse or bicycle, slower only than the steam railway train, which was confined to its tracks. The fast launch, driven at first by steam and subsequently by the internal combustion engine, became an exciting *chic* possession, 'the sportiest thing anyone could own' someone said. There was serious competition in speed between such craft, Less serious, but socially beguiling for those who were young while the old century was running out, was to go for a breathless spin in a steam launch, the ladies veiled, well muffled and tensed with the thrill of this tremendous new experience of speed, while the gay blue waters of

Right

The new form of sea speed achieved by the Union of the light steam engine and a long narrow light hull, *Niagra IV*, 111 ft in length and with only 12 ft 4 in beam. Yachts such as this were able to reach speeds of more than 30 knots, and in the early years of the present century were often used for racing. Accommodation on board was little and for day use only, the toothpick type of hull being mainly occupied with the triple-expansion steam engine and water-tube boiler.

Left

Originally *Philante*, subsequently the Danish royal yacht *Norge*, represented the highest development in design of the largest diesel-propelled yachts of the 1930s. Though in this decade the last of the clipper-style yachts in the tradition of *Corsair* (see opposite) were still being built, the new idiom of design represented by *Philante* had conquered.

Long Island Sound, or of the Mediterranean off Cannes or Monte Carlo, rushed under them at 20 knots and more.

It happens that the speed of these craft was often expressed in miles per hour. Partly it was no doubt because this produced larger numbers than knots. Also, these craft were bringing a new type of person afloat, mechanics and engineers with little knowledge of seafaring who alone could tame such temperamental exciting creations.

Nathaniel Herreshoff produced a number of fast steam launches. They were particularly popular in the New York area, where before the day of the motor car the business man could commute from his suburban home by launch, landing at a jetty perhaps a few blocks from his skyscraping city office. As early as 1887 an 85 ft Herreshoff launch made the passage from Newport to 24th Street New York, a distance of 170 miles, at an average speed of 24 miles an hour, or 20.84 knots. In 1902 Herreshoff produced the *Arrow*, 130 ft long but a mere 12 ft 6 in in beam and very light. Her speed was claimed to be 39 knots. Craft of this kind became known as toothpicks on account of their proportions. They did not skim or plane in the manner of later high speed craft, but with their narrowness and lightness cut through the water without lifting.

One of the earliest internal combustion engines to be put afloat belonged to Gottlieb Daimler, not yet famous for his cars. A 1½-horsepower motor was installed in a small launch. By 1900 many engineers in several countries, especially in France and Italy, had produced reliable automobile engines which they wished to fit into boats. Nathaniel Herreshoff still believed he could produce a small light steam plant that could drive a boat faster than any internal combustion engine, and in an attempt to prove it he built a beautiful little steam launch, *Swift Sure*. But she was beaten in a race with one of the new motor-driven boats, named *Vingt-et-Un II*, which had a 75-horsepower Panhard engine. The steam engine was defeated through its inability to compete with the high power–weight ratio attainable with the internal combusion engine.

In 1903 Sir Alfred Harmsworth, the London newspaper proprietor later to become Lord Northcliff, offered a cup, at first called the Harmsworth Trophy, though it later became better known as the British International Trophy, for international motor boat racing. The races were to be between boats of less than 40 ft in length and the engines had to be made in the country that the boat represented. This trophy was one of the greatest incentives to the development of fast motor boats.

Late in 1903 a new boat was put into the water, a narrow slip of a craft, 35 ft long, driven by a 75-horsepower Napier engine. She was named *Napier Minor*. In great secrecy before dawn one morning she was given her first trials on a reach of the upper Thames. As the sun came through the mist of the breaking day she was being hurled over the smooth water, while her engine's roar broke the quiet of the river banks and fields beyond. Shortly afterwards she won the Harmsworth Trophy at a speed of 23½ miles an hour.

The U.S.A. challenged for and won the Harmsworth Trophy in 1907, the boat *Dixie* representing the Motor Boat Club of America. To defend the trophy in the following year Clinton Crane designed *Dixie II* and took the sporting risk of guaranteeing the owner of the boat a speed of 35 miles an hour; if this was not achieved the owner would be relieved from paying for the hull or engine.

Clinton Crane produced a boat of the maximum length allowed – 40 ft – and gave her the thinnest and lightest shaving of a hull, 5 ft in breadth and weighing under 1,000 lb, which was less than half the weight of the engine installed. *Dixie II* was timed on the measured mile at a speed of 36.6 miles an hour.

Opposite

On board Gordon Bennett's steam yacht *Namouna* in Venetian waters, 1890. Painting by Julius L. Stewart.

Overleaf

Dinner afloat at Cowes in 1909 on board Cornelius Vanderbilt's steam yacht *North Star*. Painting by Jacques Émile Blanche.

Steam yacht *Zaza* in the Bay of Naples, painting by Roberto.

Nearly double the speed of the fastest sailing ship had been reached. Shortly afterwards a new boat was built in England by S. E. Saunders of Cowes for the Duke of Westminster. *Ursula* was larger than *Dixie II* but with the same kind of long, slender hull, some of the sections of which were almost semi-circular, while the hull was low in the water and shaped to reduce wind resistance to the minimum. *Ursula* was an early example of streamlining afloat. She was both fast and relatively seaworthy and racing at Monte Carlo she reached 40.3 miles an hour, or 35 knots.

But a limit of speed was being approached for this type of hull whose shape precluded the generation of lift – a degree of planing, or skimming. Further advance in speed made planing boats necessary. Such an untried and novel type of craft, originally described as a hydroplane, was initially dangerous pending the solution of the complex hydrodynamics problems involved in the planing action. The future lay with planing boats, which became commonplace in the latter half of the century; but even so far-seeing an architect as Clinton Crane soon gave up designing hydroplanes, being unable, he said, to see any way of making them safe and useful.

But by 1912 the hydroplane *Maple Leaf IV* had won the Harmsworth Trophy and attained a speed of nearly 50 knots. There are still people who can recall her lying at her moorings in Cowes harbour. Her crew would row out to her, the engines be started, and after their initial warming roar the boat would slip out into the Solent and, increasing speed, rise and skim over the surface with all the apparent ease and grace that the planing craft when running steadily is able to reveal. A new form of motion over the sea had been devised.

In spheres less specialized the internal combustion engine revolutionized the yachting scene and the techniques of yachtsmen. Motor launches became the busy attendants at the yacht anchorages and soon the yacht, other than pure class racer, without an auxiliary became a rarity. There were those who foresaw a wholesale decline in the practice of seamanship, and the modes of seamanship were certainly to be much changed by the internal combustion engine afloat.

In 1914 the 'Great War' was still the Napoleonic war. Yachting in the modern sense was the growth of the years between that Great War and the greater one on the doorstep. At this point some figures may paint a more interesting picture than words. It must be remembered that figures of yacht tonnages and numbers cannot be exact; furthermore, that they exclude numerous unrecorded small craft which though sailed for pleasure had not then attained the dignity of yacht status – a status that has followed a course of ever-diminishing size into our own day.

The total world tonnage of yachts, including steam yachts and those with oil-fuelled auxiliaries in the later years, grew thus during the period 1850–1912, see table, left.

	gross tons
1850	22,141
1864	39,485
1875	89,420
1891	206,184
1901	271,576
1912	373,575

In the sixty-two years the growth in tonnage is some seventeen fold. During the quarter century following 1850 the growth in tonnage was some four fold; in the next quarter century three fold. In the first twelve years of the twentieth century the world tonnage increased by more than twenty-seven per cent. Between 1890 and the war the growth in world yacht tonnage was due mainly to expansion outside Britain. Thus between 1891 and 1912 the increase in the total number of British yachts was nineteen per cent, while in the rest of the world it was fifty-two per cent.

By 1891 Britain had reached the high water mark of her yachting strength in relation to the rest of the world. Thereafter the proportion of yachts owned by other nations steadily increased, from thirty-six per cent in 1891 to fifty-two per

cent in 1912. In 1901 there were 470 yachts in Germany; 900 by 1912. In the same period the total of British yachts hardly changed. It is significant that the number of sailing yachts in the U.K. – that is, yachts with no mechanical propulsion – decreased between 1904 and 1912, while there was a growth in steam yachts of almost the same number. This was because many sailing yachts in this period were fitted with oil engine auxiliaries, which were classified as steam. Such figures reveal the great change overtaking the sport. But while, all over the world, there was so great an increase in the number of yachts fitted with auxiliary power, still in 1912 there were numerous without engines.

One might wish that the last Cowes Week before World War I could have been a blaze of maritime extravagance and splendour afloat, a last fling at the eager egalitarian world already gathering with such power. Actually, Cowes Week in 1913 was a most muted affair. Not only the Big Class but the several next in size made no respectable showing. The *Britannia*, in this unexacting sphere, won in the handicap class for her size. Elsewhere in Britain the racing season proceeded much as usual. The author's great-grandfather, in a letter from his yacht in the Menai strait – a mere 30-tonner with a crew of three – gave his opinion that yachting under sail was growing in popularity only among the 'lower classes'. 'It will be all power soon', he said, echoing the opinion of about thirty years earlier.

The next year the coloured programmes were in their stacks for 1914's Cowes Week, and were about to be distributed when it became evident that something bigger was about to roll over that old Cowes – but not kill it. There was no regatta. But divided from all this by the Atlantic, sailing continued for a few years in the U.S.A. with advantage to the art. The first great period of yachting, which had lain between the Napoleonic wars and that of 1914, was over.

In 1912 *Maple Leaf IV* made a speed of nearly 50 knots. Powered with an internal combustion engine and with a hull designed to lift and skim, craft such as this were able to better the speed of the long narrow steam launches of the kind represented by *Niagara IV* on p. 151.

Part III The years between 1919-1939

10 Re-establishment of the Big Class

In England the first post-war yachting season of 1919 was rather like a ghost in a world whose modes of life had changed. During that summer there seemed little promise that the years to come would see anything more than the palest resemblance of those pre-1914 summers before the holocaust. It appeared certain that first-class racing in large cutters had been killed by the war. It transpired, however, that another war would be needed to lay it out finally. The interval between the two world wars produced the last – presumably for ever – of Big Class yacht racing. For the space of rather less than two decades a Big Class of the utmost distinction came into being again in Britain and the U.S.A. – the only two nations in the history of yachting to raise such yachts.

The initial impetus came from an announcement by King George V that he intended to fit out his cutter *Britannia*, now in her twenty-eighth summer, the one survivor from the great seasons of 1893–6. There were in existence a few of the pre-war 23-Metres to the yet new International rule; *Nyria* (1906), *Brynhild* and *White Heather* (1907), while a new boat to the class was on the stocks in the re-establishment of Big Class racing; this was *Terpsichore*, later and better known as *Lulworth*. In this year the International Rule of 1907, which had been due to expire in the course of the war, was replaced by the Second International Rule, differing in important respects from the former, the most important of which lay in the measurement of sail area. It was a change that marked a fundamental difference between the racing yachts of all classes prior to 1914 and after 1920; it was a crucial step towards modern yachting. Where the earlier yachts had stepped out under clouds of canvas, the rigs of each new season spreading wider and finer than those of the last, the sail areas carried by yachts were now suddenly restrained, and also the size of the crew to handle them. In this respect some of the extravagance was going out of even the most abundantly provided racing yachts of all.

The Big Class of 1920 and the next few years was essentially a menagerie class. With the 23-Metres and the old 1890s rater *Britannia* was the Herreshoff schooner *Westward*, carrying forward into such a different era the great days of schooner racing. There was the lovely 93-ton cutter *Moonbeam*, a new yacht and a cruiser-racer in the splendid manner rather than a dedicated racer, which won the King's Cup at Cowes in 1920. In the course of this year's racing fourteen yachts raced in the Big Handicap class at one time or another, ranging in size from the 338 tons of *Westward* down to 80 tons, and including schooners, yawls and ketches as well as cutters, the heterogeneous fleet held together by a not very satisfactory handicapping system. The hard racing core of the fleet, and the most regular starters, were *Britannia*, *White Heather* and *Nyria*, while the new *Terpsichore*, which should have belonged with this trio, proved initially a disappointment. Thus it was in 1922 and 1923, though *White Heather* had fallen out temporarily while *Terpsichore* had gained some form.

Meanwhile *Nyria* assured her, and her owner's, place in yachting history by becoming the first big cutter to assume the Bermudian rig. It was a brave move to which her owner, Mrs R. E. Workman, had been persuaded by Charles Nicholson, whose technical daring had been shown in the America's Cup challenger *Shamrock IV* despite her defeat in 1920 (see below). Marconi was much on people's minds at this time, and *Nyria*'s new skyscraping mast, initially kept aloft by a mass of wires led over an ugly multiplicity of spreaders and struts, could only suggest a radio transmission mast rather than that of a sober hard-racing cutter; so the rig was called 'marconi' and widely damned, though not the popular owner of it, who was no less widely admired. *Nyria* was over the threshold

The first Big Class yacht to carry the Bermudian rig, *Nyria* in 1921 raced against gaff-rigged competitors. When first rerigged the staying of the tall mast was more complex than that illustrated here, which is a clean arrangement of three spreaders and a single jumper strut. *Nyria* had also been a pioneer when she came out in the first decade of the century in having a hull construction in accordance with Lloyd's rules for the highest class, thereby terminating the day of excessively light construction of racing yachts.

into the new age, and no more large racing cutters of the first class were henceforward to be other than Bermudian rigged. Aerodynamics were creeping into yacht racing. The Bermudian rig, capable of producing more thrust per square foot of area than the gaff rig with its flouncing jackyard topsail, further advanced the cause of smaller, more efficient sail plans, which were earnestly demanded in this new world where the supply of paid hands 'racing every day of the week, every day but Sunday' was running as short as was the money to pay them their justly increased wages.

Indeed, it was doubtful whether racing in the largest class of yachts was going to survive. By the season of 1923 there were only three well matched big yachts racing regularly, and still, at the end of the 1927 season, a number of boats ranging between three and five uncertain starters could not give confidence in the future. The Second International Rule of 1920 was intended to apply only to yachts rating under $14\frac{1}{2}$-Metres. A slightly modified version of this rule was introduced for application to larger yachts. It had a brief existence, but produced two new 23-Metres for the 1928 season, *Cambria* by Fife and *Astra* by Nicholson, and a third, *Candida*, for 1929. All three were the Bermudian-rigged creatures of the new age, with fine qualities inherited from the old.

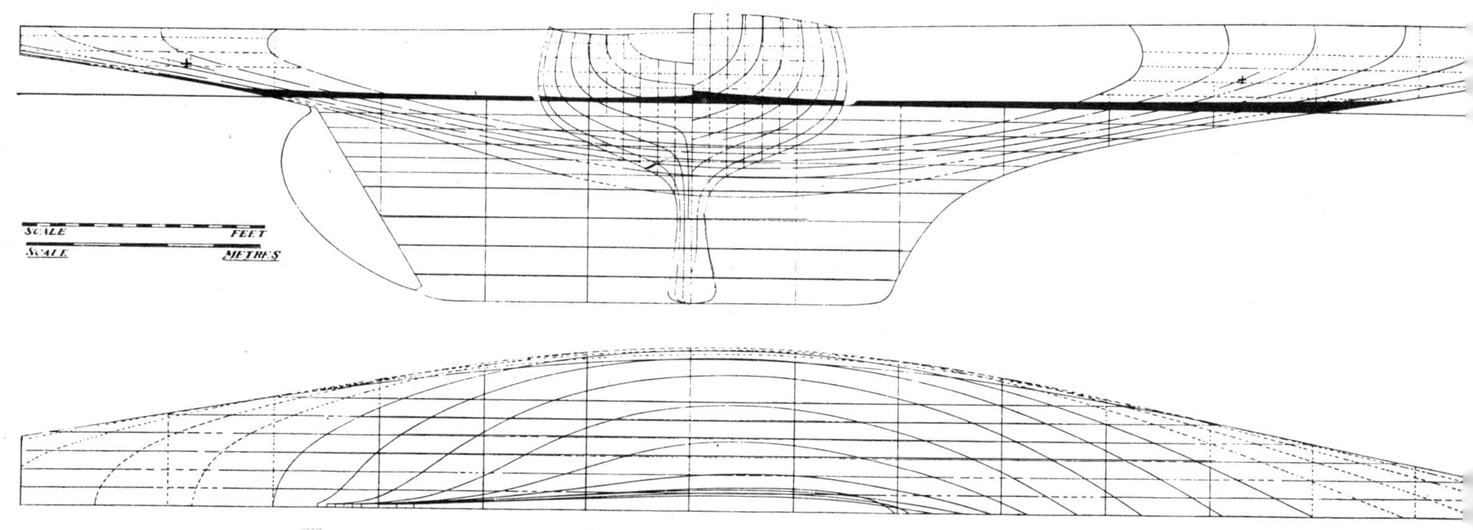

Above and opposite

By gaining the first two races for the America's Cup in 1920, *Shamrock IV* needed only one more race to win. In the design of his first challenger Charles E. Nicholson produced a bulb-keel skimming dish of an extreme form and elaborately unusual construction, with mainly triple-skin wood planking and longitudinal framing, reinforced by widely spaced steel-web transverse frames. Though the two headsail rig was often carried, a single headsail with a club along its foot was set sometimes, as may be seen in the photograph. The topmast was socketed into the lower.

Prior to these events another America's Cup series had been sailed. The challenge had been initiated in 1913, and the remarkable challenger *Shamrock IV* was on her transatlantic crossing in 1914 when war was declared. She was laid up in Brooklyn and, not surprisingly, neglected for the next six years; then taken from drydock for a resumption of the challenge in 1920.

Shamrock IV, the first challenger designed by Charles E. Nicholson, now the leading British designer, was the ugliest yacht yet to have left England in quest of the cup; also the most technically advanced and extreme racing machine. Nicholson had contrived to force an immense scow type of hull through the terms of the American Universal Rule. The graceless snub-ended, hog-sheered hull was probably the most lightly constructed of its size ever to have been produced; the triple-skinned wood planking laid over a complex structure of navaltum web frames, steel intermediate frames and silver spruce stringers, represented the ultimate sophistication of composite shipbuilding. Inside, the hull was virtually gutted from end to end, all plumbing removed for racing, there not even being facilities in this great yacht for providing cups of tea – one of the many circumstances that produced a crew, still no doubt war weary, that was in a state verging on mutiny throughout the contest. This light, lead-loaded shell of a hull, of great beam and little depth, was immensely powerful. Charles Nicholson had

adopted wholeheartedly the design philosophy of Nathaniel Herreshoff; in 1920 this old master, producing the last of his six successful defenders came nearer to defeat than he had ever done before, at the hands of a designer producing his first challenger.

For the first time in cup races both challenger and defender were steered by amateurs; a pointer, not unrecognized at the time, to the spirit of the budding post-1918 era. At the helm of the defender, *Resolute*, was the Secretary of the U.S. Navy, Charles Francis Adams, whose habit of steering the yacht in a Panama hat and braces inspired the photographer Morris Rosenfeld (whose work is represented in this book) to obtain a close-up shot of the new style of helmsman. He succeeded in manœuvring his launch close enough and was not discouraged when Adams waved his hat at him and shouted 'Rosy, get to hell out of it'. At the helm of the challenger was Sir William Burton, then Vice President of the Yacht Racing Association, with Lady Burton as timekeeper, a fact that inevitably stirred the maritime prejudices of the paid hands.

The long stay in drydock had led to deterioration in *Shamrock*'s sophisticated composite structure. This could only add to the delicate state of the crew's already fragile morale, and fears – which were without foundation – that the yacht would break up under them, which did not improve their state of mind when the shallow hull pounded violently in a seaway. Relationships among the afterguard were unhappy also, though for different reasons. Charles Nicholson, the designer and also a practical yachtsman of skill, and Sir William Burton were in frequent disagreement, and in the course of the five races which comprised this close contest *Shamrock*'s afterguard suffered various changes as one or another of them was left ashore in order to soothe the friction. And this in turn reacted on the crew: ''ell of a note, three skippers' one of them confided to Sherman Hoyt, the popular representative of the New York Yacht Club on board the challenger.

Shamrock IV proceeded, however, to win the first two races, which made it necessary for the defender to win the next three in succession or lose the cup, the continued display of which in the New York Yacht Club becoming more open to doubt than ever before in the story of the contests. It had been demonstrated that there was not much to choose between the speed of the boats in light winds; *Resolute* was able to sail higher to windward than the challenger, but under these conditions the latter footed faster through the water. In harder winds the great power of the challenger's hull was an advantage, but a weak-socketed topmast offset this.

The racing in general was not stirring despite the closely matched qualities of the boats. The first race went to *Shamrock IV* owing to the disablement and retirement of *Resolute*, who had trouble with her main halyard. But *Shamrock* was in a state of near disablement also with headstays slackening and slopping out to leeward and the masthead bending aft. Both yachts had run through a heavy squall and *Resolute* had worked out a lead of just short of five minutes at the first mark. She was the smaller boat, with less sail area than the challenger, so with the time she was allowed her victory at this point seemed assured. But after she had retired *Shamrock*, also on the verge of retiring, was nursed over the line to a victory that did nothing to restore morale among her crew. But she won the second race in a confident style. Sir William Burton, whose handling in the first race had produced criticisms enough to raise a rumour that he was to be displaced, showed his true form as one of the few brilliant big yacht amateur helmsmen of the day.

When in the third race, which would bring *Shamrock* the cup could she win it, the challenger gained a comfortable lead, Sherman Hoyt is reported to have said to Charles Nicholson 'that cup's about to cross the pond'. But though *Shamrock* crossed the line first she failed to cover the time allowance she owed to *Resolute*, and the latter won on corrected time. But still there were two races to be sailed, a victory in either of which by *Shamrock* would win the contest. But such was not to be. The fourth race went to *Resolute*. On board *Shamrock* there had been constant worry about the topmast, and a new one had been fitted before this race. When it proved no better than the original, it was Charles Nicholson's turn to make a prophecy to Sherman Hoyt: 'Your Cup is safe.' And so it was to be. Two attempts were made to sail the fifth race before, on the third attempt, it was completed. Never before in the history of America's Cup racing had the yachts started what must be the final race each with two wins. Never perhaps had the yachts been so well matched. It can only be regretted that luck ruled on this day; but any regret must be well tempered with philosophical resignation; for all who sail are aware that they offer much to clodhopper fortune. During a race of fickle winds the clodhopper stepped casually over *Shamrock*'s chances.

Until 1925 there was no central authority for yachting in North America comparable with the British Yacht Racing Association and other European bodies. The U.S.A. was thus not represented on the International Yacht Racing Union. This absence of a national body until so late in the yachting story brought with it a lack of uniformity in sailing and measurement rules which was, however, less disorganizing for the sport of yacht racing than it would have been amongst the smaller European nations with their relatively tiny lengths of coastline and more intimate yachting communities. But the need for a controlling body was becoming evident.

The North American Yacht Racing Union (N.A.Y.R.U.) was founded at a meeting of United States and Canadian yachtsmen in New York in November 1925. A leading figure in its creation, and the first President, was Clifford D. Mallory, who had been an enthusiastic racing helmsman since the 1890s, and was well known in the yachting circles of Canada as well as the U.S.A. The constitution was drafted by William A. W. Stewart, a lawyer who was later Commodore of the New York Yacht Club. The object of the Union was stated as 'to encourage and promote the racing of sailing yachts and to unify the rules'. It could lead but not command, its rules binding nobody. There were three committees, for the executive, racing rules and measurement. The body went into action in 1926.

Now for the first time the U.S.A. had a yachting organization through which it could act in international affairs. But it was not until a quarter of a century later that the N.A.Y.R.U. became affiliated with the I.Y.R.U. in 1952; and still it retained its own racing rules, differing in important respects from those of the European body.

The course of events was tidier in regard to the yacht measurement rules. Early in 1927 a committee of the N.A.Y.R.U. and I.Y.R.U. met in London, and the outcome was the adoption by the former of the International Rule of Measurement which governed the 6-, 8-, 10- and 12-Metre classes throughout Europe. In the U.S.A. there were then two rules of measurement governing first-class racing yachts, the Universal rule, which had been in use for many years and to which no English or European yachts had been built, and now the International rule which could bring competition between the two sides of the Atlantic. At the 1927 international conference, which was attended by Great Britain, France, Holland and Scandinavia, as well as the North American representatives, certain alterations were made to the International rule to bring it more into line with American practices, and in a report of the meeting by Clinton H. Crane, in the *Boston Evening Transcript*, he said: 'It seems to me an extraordinary thing that we should be able to go to London and criticize a rule which had been developed abroad and not be met with immediate antagonism.'

It was a dawn of close transatlantic co-operation in yacht racing matters. While not members of the I.Y.R.U., American representatives of the N.A.Y.R.U. and New York Yacht Club were enabled to attend meetings of the I.Y.R.U. as observers without voting rights.

Charles E. Nicholson at the helm of *Candida*. She is up to weather of *Britannia*, with *Lulworth* under *Britannia*'s lee and *Cambria* astern.

In 1928 the Yacht Racing Association took a course which within a few years was to prove inconvenient. The Association adopted a rule that was essentially the International rule with very minor adjustments for yachts above the size of the 12-Metre class. In the revival of Big Class racing, which was to lead to the last phase of this kind of lofty style in yachting, we have seen that three new yachts were built to the 1928 rule, *Astra*, *Candida* and *Cambria* rating 23-Metres. In the same year, 1929, Sir Thomas Lipton had his last challenge for the America's

Cup issued by the Royal Ulster Yacht Club. America's Cup yachts were governed by the American Universal rule's largest class, the J-class of 76 ft rating. The Americans proceeded to make the strongest defence of the Cup yet mounted, and four of the large J-class sloops were soon under construction. The unfortunate situation had therefore arisen that while these four yachts, and one in England – the challenger *Shamrock V* – were designed to one rule, Britain was tied to another rule to which three fine yachts had already been built.

The two rules, though quite different in form, produced yachts remarkably similar in type. In 1930 the two nations agreed that for yachts above 14½-Metres rating both should use the Universal rule, and Britain thereupon discarded the short-lived 1928 rule. The Universal rule, as applied to the J-class, had added to it some further safeguards. The America's Cup contest of 1930 found the British challenger shocked at the American defender's virtually gutted accommodation. This led to requirements – which were accepted – that rules should insist on the proper fitting of living accommodation; also that there should be a limit upon the lightness of masts, another respect in which the 1930's cup defender had sprung a nasty surprise upon the challengers.

There were those who criticized the British acceptance of the American rule, claiming that the yachts it produced were less suitable than the three Graces, *Cambria*, *Candida* and *Astra* which had just come from the builders. But there is no reason to believe that under the pressure of America's Cup racing during the 1930s this breed of yacht would not have become, as did the Universal rule yachts, dedicated, extreme racing machines; while the merits of a common rule between the two nations were unquestionable.

As Big Class yacht racing swung into its last phase, with behind it all the technical expertise of the twentieth century, the inherent conflict of ideas behind this type of vessel became growingly obvious. There had always been a belief – more precisely a vague and inarticulate feeling – that a racing yacht, except of the smallest size having no accommodation, is ethically unacceptable upon the face of the waters unless she is something more as well. Speed is an Attic quality, and ships in all ages have been loved for possessing it, while devoid of all sterner virtues. But just to be fast is not enough; that is to be a racing machine and repulsive to seamanlike tastes. A yacht big enough to be lived on board should have all the amenities providing comfort for this purpose, which unfortunately entails weight that reduces stability and speed. In this dilemma yachtsmen and their legislators found themselves alternately getting themselves as fast yachts as possible and then making rules to prevent themselves doing so. Herein lies no small part of the history of yacht racing.

It was as though yachtsmen and legislators in unison were raising the old prayer, 'Make us good – but not just yet.' The last big cutters on both sides of the Atlantic were governed by rules enforcing minimum standards of accommodation; but having assured this the hard racing owners refrained from living in their habitable hulls, and in the cause of lightness accommodation was skimmed to the maximum extent that could be twisted from the rules; while the gracious living was conducted next door in the motor yachts that attended the racing fleets. Thus it was in the J-class yachts, which were virtually day racing craft, though of some 130 ft overall and displacing about 160 tons. In the much smaller 12-metre class there was also space for comfortably habitable accommodation; but no owner could hope to win in such a hotly contested class if his boat carried the burden of a small household afloat. These large inshore racing yachts of the 1930s were the last stage of development in one direction;

racing machines they may have been, but with a superb performance remembered by a dwindling few, and by them regretted.

The Americans rationalized the situation more readily than the British, and this became startlingly evident when *Enterprise*, the cup defender for 1930, became open to view (see below). Inside the ineffably graceful hull of adequate strength were found winches, coiling drums and lockers for sails, where the cushioned settees of the saloon, the glass-surrounded dressing-tables of the staterooms were expected to be; while the hull lining and sole were light wooden slats.

In one respect, however, the ingenuity of designers now had a curb put upon it which was rationally observed. Very light construction in large racing cutters, and to a lesser degree in the smaller also, had too often reached absurd, if technically brilliant, limits. *Shamrock IV* had been an example of this, if less extreme than *Independence*, a yacht built as a prospective America's Cup defender in 1901. She had a bronze plated hull with nickel-steel frames intricately tied together by a mesh of tubular struts, braces and tie rods. The deck was of aluminium. No one worried about mixing metals three-quarters of a century ago; electrolysis would have finished *Independence*'s life quickly had she not been broken up a few months after first setting sail.

Her end almost did come within a few days of her being launched, when she was being towed through a short steep sea at a speed only a little in excess of a knot. Tie rods began snapping like pipe stems, tubular uprights buckled, plating sprang and pulled its rivets, and the yacht took in water at a rate that made

Advanced engineering in yacht design as seen in *Enterprise*, which defeated *Shamrock V*, with the main accommodational spaces packed with winches and spooling drums and the flooring of light wooden laths. The hull plating and framing was of aluminium alloy.

foundering seem imminent; men stood by the dinghies on deck ready to launch them instantly. On this occasion she was saved, but she was never afterwards other than sorry for herself.

Repairs, so far as possible, were made; but while many parts were strengthened nothing could be done to remedy the basic fragility of the hull – 140 ft long with shallow champagne-glass sections below which a slender fin plunged to a draught of 20 ft, carrying at its base 80 tons of lead representing about sixty per cent of the yacht's total weight. During *Independence*'s brief racing career the most vital piece of equipment on board was a pump capable of delivering 4,000 gallons of water an hour; and it was kept running for twenty minutes every two hours. Race, however, she did for a short time, under a sail area of 16,000 sq ft – as much as has ever been set from a single mast in the history of sail. To have been on board *Independence* during a race must have been a sailing experience to darken a lifetime. Even her owner, Thomas W. Lawson, who thought that something might have been made out of this flouncing freak of a yacht, wrote: 'When she lay down, the braces to leeward buckled and those to windward straightened out; on the other tack those that had buckled straightened, and others bent into crescents.'

Independence was a memory of nearly three decades ago when the step was taken in 1928 of ruling that in future races for the America's Cup the yachts should be built to Lloyd's scantlings and hold the Society's classification certificate. In 1908 the International Yacht Racing Union had adopted rules drawn up by *Lloyd's Register* for the other international classes. Now the principle was extended to the biggest racing yachts. This assured that the materials, the disposition and size of the structural members, should be in accordance with conventional, and indeed rather unadventurous practice based on experience. The racing yachts in all sizes of the international classes thus became of adequate strength and able to last a racing lifetime. Some freedom had been taken from the designers, but the awful example of *Independence* could not be overlooked.

There is little logic in the the evolution of racing yachts of the more expensive kind. Now, in 1928, adequate strength of hull had been assured. Further, the great sail areas of the past had been abolished – again in all classes. While *Independence* had spread 16,000 sq ft of canvas and her whole wardrobe of sails had weighed 7 tons, the new breed of big cutter carried a mere 7,500 sq ft in the aerodynamically more efficient Bermudian rig. But dedicated racing people can make a farce of their own technical progress. It had now been assured that hulls should be of adequate strength, but no scantling rules governed the strength of rigs. Here designers could lavish ingenuity on the high Bermudian rig, applying to it engineering principles that could assure increasing lightness though not strength for all occasions. The old gaff-rigged yachts with solid wooden lower masts and fitted topmasts frequently sprung their masts and lost their spars. They could indeed sustain much damage while retaining the basic rig with the ship. But with the evolving Bermudian rig, highly stressed and cunningly engineered, the failure of a quite small member of the rigging or associated fittings could bring the mast and whole rig down in grand collapse. This was to happen frequently in the days to come.

In the spring of 1929 the above challenge for the America's Cup, the first in thirteen years, and of the post-war period, was received by the New York Yacht Club from the Royal Ulster Yacht Club, on behalf of the same Sir Thomas Lipton who had made the last four challenges. And in this year Starling

Burgess's childhood ambition was achieved. He was able to emulate his father in designing America's Cup defenders. On the arrival of Sir Thomas Lipton's fifth challenge he received, together with three other architects, a commission for a J-class cutter to the Universal rule, the first of their kind to be produced.

Burgess's commission came from a syndicate headed by Winthrop W. Aldrich, Vice Commodore of the New York Yacht Club; and also on the syndicate was Harold S. Vanderbilt, of the family as well known in national as yachting life. The success of Burgess's defenders three times during the next eight years owed much to this remarkable helmsman, who was also possessed of a keen brain and great powers of organization. Vanderbilt was appointed helmsman of the syndicate yacht. Another son of the former generation of yacht architects produced a potential defender in 1930, Francis Herreshoff, son of Nathaniel. His *Whirlwind* was, for several reasons unconnected with her basic design, outclassed from start to finish by the other three boats produced for the defence by Burgess, Crane and Frank Paine, son of the General Paine who had sponsored Burgess's father. But in two respects of basic design the years proved *Whirlwind* to be the most advanced boat of the new fleet, for in her great length, up to the limit allowed, and in her double headsail rig, she anticipated the last of the J-boats. And the boat produced by Burgess, the smallest of the fleet, was in this respect the most wrong of the lot; including even the unsuccessful challenger by Nicholson. She owed her success to other advanced qualities.

The rules governing the contest had been changed in most important respects since the last challenge and presented designers with a new set of problems. It

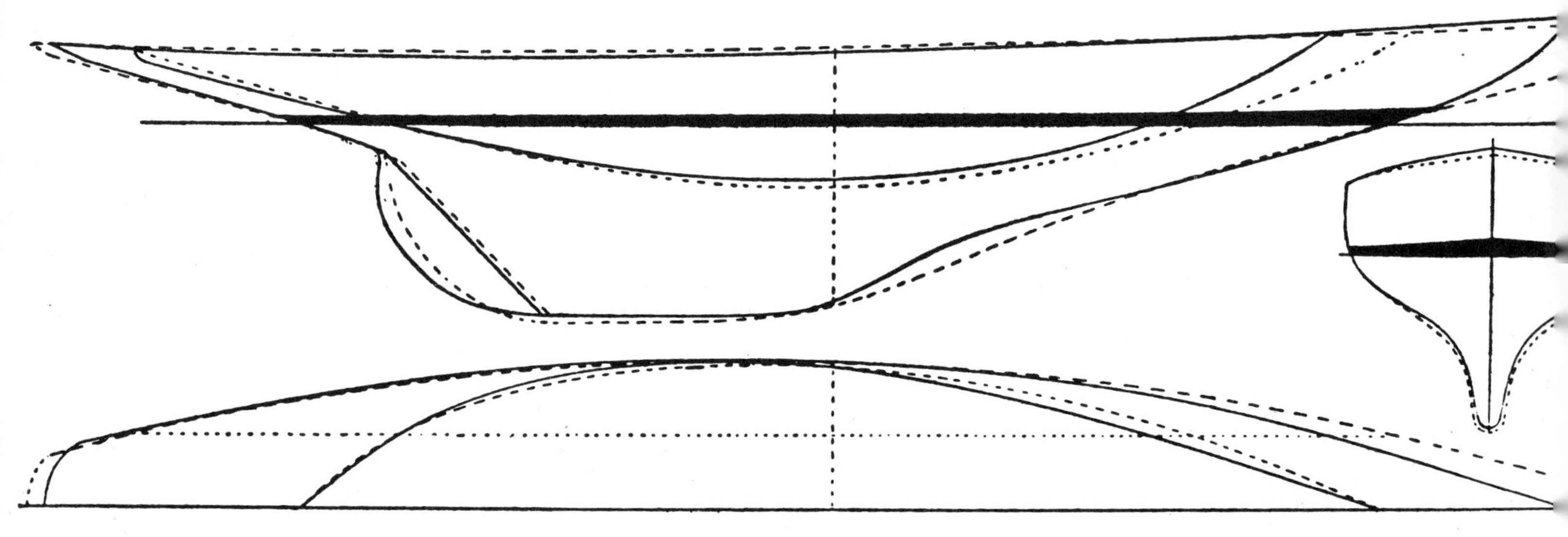

was specified that the rating of the challenger and defender should not exceed 87 ft under the Universal rule. Formerly a limit had been set upon waterline length, which meant that challenger and defender were likely to be similar in waterline length but might vary quite widely in sail area. But the Universal rule, by which the rating was now assessed, virtually fixed sail area while allowing a great range of lengths and displacements. Increase in length, according to the rule, made an increase in displacement necessary, a certain minimum displacement being stipulated for each length. The limits were approximately thus:

Length waterline	Displacement	Sail area
75 ft	108.4 tons	7,614 sq ft
87 ft	163.8 tons	7,546 sq ft

Opposite above

Designer and helmsman on the deck of *Enterprise*. Starling Burgess (left) like his father Edward Burgess was responsible for the design of three America's Cup defenders. Harold Vanderbilt, who sailed these three defenders, was one of the finest racing helmsmen of his day. His name became associated with changes in the code of yacht racing rules.

Below

The lines of *Enterprise* drawn by Starling Burgess. The skeleton of these lines are also shown opposite superimposed on those of *Britannia*. Thirty-seven years separated the appearance of the two yachts, but it will be seen that an ideal of beauty represented by *Britannia* (full lines) was closely echoed in *Enterprise*. But despite the comparatively small difference in the shape of the two yachts, *Enterprise* was much the faster, mainly owing to the engineering refinements of her hull and rig.

Yachts with the above displacements would both rate 76 ft, which meant, rather surprisingly, that one yacht might have some fifteen per cent more length and fifty per cent more displacement than the other yet would have to carry almost the same sail area to rate at the same figure of 76 ft. Here was a remarkably wide range of lengths within which designers had to select the one most suitable.

There was some guide from past experience. Racing experience had shown that, in the smaller classes to the same rule, it paid to take the greatest length allowed and to accept the greater weight which then had to be driven by almost the same sail as allowed to shorter, lighter boats. The reason for this was not clear until a few years later, when an accurate method of tank testing models had been evolved. And there was also doubt whether the experience with the smaller boats could be applied to the much larger J-class. Was there not a danger that with such big craft, the maximum length would entail a boat too heavy to be driven by the sail area allowed under the normal racing conditions? This was the designers' problem. Starling Burgess set about its solution with the greatest care and forethought, using the best hydrodynamic skills available; but he arrived at an answer wider of the mark than any of the other competing architects.

Meteorological records of the United States Weather Bureau for the preceding twenty years were examined to determine the likely wind strengths over the newly chosen cup course, an ocean course laid south of the Brenton Reef Light vessel off Newport. The conclusion reached was that for the anticipated wind strengths, the new yacht should be designed to give her best performance at a

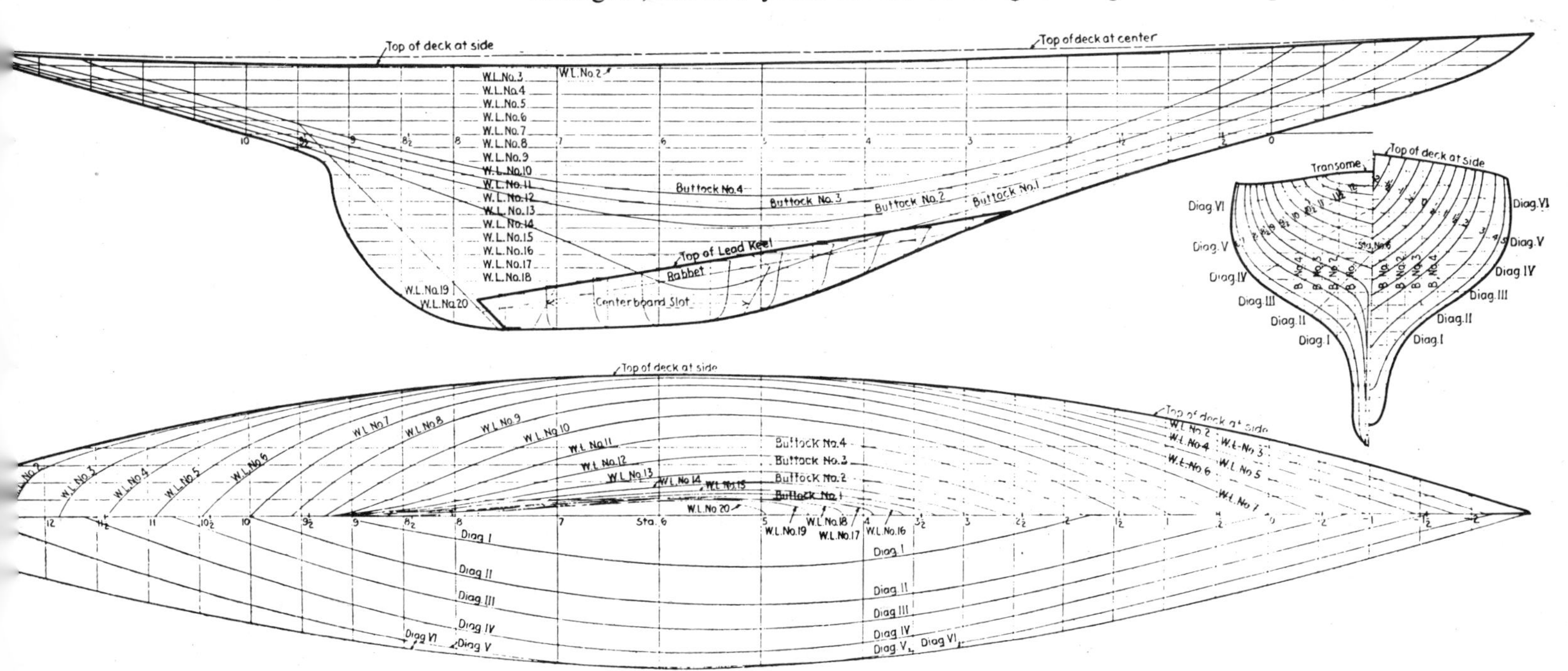

speed of $10\frac{3}{4}$ knots. It will be evident that to determine this figure required some judgement that could not be supported with assured data.

Harold Vanderbilt had favoured a hull form resembling that of his M-class sloop *Prestige*, designed for him to the same rule by Burgess some years earlier. The hull form now produced by Burgess was notably different, but was the one unanimously chosen by the owning syndicate when presented to them as a half model. The hull had considerable beam, a stiff midship section, this being considered necessary by Burgess to assure adequate stability, while the wide-vee bow sections, though considered liable to pound (which in practice they proved to do) were believed to give slightly higher speed in light winds.

Shamrock V at Gosport, England, a few moments after her launch at the yard of Camper and Nicholsons. She was conventional in type, construction and rig, unlike *Shamrock IV*; and in 1930 suffered a convincing defeat in the America's Cup contest at the hands of a more advanced yacht.

At the same time, Nicholson at Gosport in England was creating the lines that were to be *Shamrock V*, and Crane, Herreshoff and Paine were working out their new designs, each to dimensions and hull forms unknown to the other. Crane tells a rather sour story of this period. In the New York Yacht Club one afternoon Burgess came up to him to 'show me a set of lifts which he said he had gotten from the model maker who was building the model of one of the new designs I had made for a defender. I said, "Starling, why did you bother to go to all the trouble of getting these lifts from the model maker? If you had asked me, I would have shown you the design I was making." He rather flushed at this, and said I must consider it a boyish prank. A little later on, when we built *Weetamoe* and did not use this model, Starling was distinctly annoyed.'

In the crucial choice of length, the four American and one British designs ranged between the 80 ft of *Enterprise*, the smallest, and the 87 ft of *Whirlwind*, which was the longest and to the limit allowed. The next smallest was Charles Nicholson's *Shamrock V*, but she was appreciably longer than *Enterprise* at 81.4 ft, while above her came *Weetamoe* and *Yankee* in that order. We can now be sure, though in 1930 the architects did not have the data on which to judge this, that had any architect gone to the 87 ft waterline allowed and, unlike *Whirlwind*, the boat had been properly tuned and equipped, she would have proved the fastest of the yachts.

While in Gosport the wood-planked *Shamrock V* was taking shape, her planking laid on steel frames, her wooden mast carved from spruce, and while the riggers produced the many tackles that would be used in handling her rig in the absence of winches, at the Herreshoff Manufacturing Company – no longer owned by Nathaniel but still staffed by the people trained by him – a mechanized yacht took shape, bronze plated with polished bottom, two centreboards, and an echoingly empty interior which we have seen was to be packed not with the accommodation fitted into *Shamrock V*, but with winches, spool coiling drums,

and sail lockers. This mechanical yacht was not a new design conception for the America's Cup. Nathaniel Herreshoff having shown the way a generation earlier, while Charles Nicholson had followed it with *Shamrock IV*.

The races themselves that year lacked interest. With her heavy wooden mast, compared with *Enterprise*'s mast of duralumin, *Shamrock* began with a crippling handicap – someone remarked at the time that to equalize things *Enterprise* would have needed to hoist a ton weight up to her lower spreader, for *Shamrock*'s mast weighed 6,350 lb compared with the 4,000 lb of *Enterprise*'s. The mechanization on board the defender and the superb organization of the afterguard, under the presiding genius of Vanderbilt and the eccentric genius of Sherman Hoyt, produced a combination of ship and its handling that was fated to win. But though the challenge was toppled in four consecutive races won by *Enterprise*, the British yacht was by no means disgraced; and to most people she could only appear as much the more desirable type of vessel. At her helm was Captain Ned Heard, the last professional to steer an America's Cup yacht; he was to be on board the next challenger four years later, again as professional skipper, but not at the helm.

Sir Thomas Lipton opened negotiations for a sixth challenge during the summer of 1931, the year when England went off the gold standard. The challenge was postponed as a result and never renewed, for shortly afterwards Lipton died. To the general public on two sides of the Atlantic he was the most celebrated of all yachtsmen; to yachtsmen a great publicist whose five challenges for what in America was known as the 'International Yacht Race' sweetened Anglo–American yachting relationships for the sake of selling tea, and in the process so greatly benefited yachting.

In the spring of three years later a large blue racing yacht was sent down the ways at Camper and Nicholson's Gosport yard, coming to rest on the waters of

Portsmouth harbour, where so many predecessors famed in yachting history had first found themselves afloat. The yacht held the attention of more people than those who watched her being christened *Endeavour* that day; for it was widely known that she was going to be the fifteenth challenger for the America's Cup later in the summer. As one yachting correspondent asked at the time, will she be just one more in the lengthening line of yachts that have failed to lift the cup, or will she gain a place in yachting history no less secure than that of the *America* herself? She joined the failures; but it was a failure splendid enough to be memorable.

The America's Cup was brilliantly and successfully defended in 1934 by the slower boat. Important changes had been made in the rules governing the J-class following the uncomfortably intrepid *Enterprise*. The minimum weight of masts was increased to 5,500 lb stripped; the minimum weight of cabin fittings was also raised and rules prohibited the setting-up of standing and running rigging below deck. These conditions, and others of less importance entailing more constructional weights and hence a lower proportion of ballast, outclassed *Enterprise*, which it has been seen was a short boat for her class and of comparatively small displacement. So a brilliantly contrived racing machine of almost indecent cost became obsolete after three years: 'If you drive along the waterfront in Bristol, Rhode Island, as you pass by the Herreshoff plant you will see the unburnished hull of a large sailing yacht standing on shore close by the north shop . . . Already the grass is growing here and there beneath her keel. While we leave her resting there, let us think of her proudly sailing the sparkling seas, the white

The first *Endeavour*, which in 1934 so nearly won the America's Cup. The boom, of triangular section, known as Park Avenue, will be noticed – its object being to allow the foot of the mainsail to assume an aerodynamically effective curve. This yacht is now preserved, helped by the British Maritime Trust.

ghost of the Western Ocean, "a lovely vision for any eyes that see her there tonight".' Thus wrote Harold Vanderbilt when the yacht was hardly a year old. It is not to be questioned that the last Big Class racing yachts, which were soon to disappear for ever, had become in their last stages a breed over-ripe and with a certain *fin de siècle* decadence.

Two other of the boats built for the 1930 defence were capable of being altered for the new conditions, *Yankee* and *Weetamoe*, and both were extensively rebuilt and rerigged for another year of harsh competition. They were joined by a new yacht designed by the creator of *Enterprise*. Starling Burgess had started in 1931 to prepare the design for a new defender, when in that year there was the possibility of Lipton challenging again. The yacht emerged as *Rainbow*, and along with *Yankee* and *Weetamoe* she prepared to meet the fourteenth challenger.

This yacht was the third J-boat to be built in England, for following *Shamrock V* Charles Nicholson had in 1933 produced *Velsheda* for W. L. Stephenson, who was the head of the English Woolworth chain. This was the only yacht of the class in England or America not to be built specifically for America's Cup purposes. After an unpromising start to her racing career *Velsheda* proved to be much superior to *Shamrock V*, and it had been agreed that should she prove faster than the new *Endeavour* she might be substituted as challenger – a relaxation in the terms of the contest.

Endeavour was an all-steel beauty of which the eminent English yachting writer John Scott Hughes was later to say 'The darling jade nearly broke my heart'. She was a larger boat than *Rainbow*, though not to the limits allowed by the class; and here Nicholson's concept was in advance of Burgess. The dimensions of the two yachts were:

	Rainbow	*Endeavour*
Length overall	128.56 ft	129.7 ft
Length waterline	82 ft	83.8 ft
Beam	21 ft	22 ft
Draught	14.95 ft	15 ft
Displacement	138.13 tons	143 tons

Endeavour was the larger and more powerful yacht. Experience was to prove during the next few years that under the J-class measurement rule the optimum boat should be to the maximum length allowed, 87 ft on the waterline. With *Shamrock V* and now again with *Endeavour* the British designer was in advance of the American in feeling his way towards the longest waterline permitted, which both were to reach in the boats for the 1937 contest.

While *Endeavour* was plated in steel above and below the waterline, *Rainbow* had bronze plating and steel topsides, the bronze providing, it was hoped, a better surface finish. *Endeavour*'s mast was in steel, circular and of small diameter, a form of construction Camper and Nicholson's raised to an advanced technique which might be described as the highest refinement of the boilermakers' art. *Rainbow*'s mast, in contrast, was in duralumin and of larger size fore and aft than athwartships, not for the purpose of streamlining but to give the spar the greatest possible fore-and-aft rigidity.

Sopwith's rival in aircraft manufacturing, C. R. Fairey, who joined the J-class by purchasing *Shamrock V*, had evolved in the wind tunnel a new form of jib, the double clewed or quadrilateral jib with two sheets. It proved a singularly powerful sail and Sopwith was enabled to equip *Endeavour* with one. The secret was soon out and the Americans copied the sail. They were, however, in advance of the British in the use of the sail they favoured so well, the genoa jib, which in boats the size of the J-class became of excessively large area and difficult to handle. *Rainbow*'s superiority in genoas and their handling was unfortunate for *Endeavour*. A notable development had occurred in the J-class yachts by the time the new defender and challenger were out on the water. While, in 1930, both *Enterprise* and *Shamrock V* had sailed under the three-headsail rig, now the yachts carried two headsails only, or indeed one alone when under genoa. It was an example, so often found in yachting development, of the big boats following the lead of the smaller.

It is of more significance to the social than the yachting historian that *Endeavour*'s professional crew, with a few exceptions, came out on strike for higher wages on the eve of the yacht's departure. Sopwith, who could be tough and opinionated, and was also incensed at the pistol-holding style of the mutiny, which occurring so late made it impossible for another professional crew to be engaged, would have nothing to do with the demand; whereupon Sherman Hoyt, who was to play an important part in the defence, cabled from America 'Commiserations but congratulations on your stand'. But the 'stand' was one of the small factors that contributed to the cup remaining still in its case in the New York Yacht Club after that year's contest. By a remarkable feat of organization, and with the help of the Royal Corinthian Yacht Club in Burnham-on-Crouch – a club for which the training of youth was an important role – a crew

T. O. M. Sopwith (later Sir) at the wheel of *Endeavour*.

of amateurs was quickly collected. Primarily small boat sailors, though experienced, they inevitably lacked that fine edge of training in the big boat expertise of the American crew of professional Scandinavians, who at that time were the most skilful racing-yacht hands in the business. These too, however, were creating a similar kind of problem, though less consequential, for their owners. In 1934 the syndicates running potential defenders had agreed to a uniform scale of wages and prize money for professional hands, thus reducing one source of expense in the millionaires' hobby of America's Cup racing. The crew objected. On the American side a compromise produced peace and the continued services of the professionals.

Despite the fracas *Endeavour* and the owner's motor yacht sailed for the U.S.A. on the date originally planned. After being refitted on arrival at Newport, the challenger proceeded to tune-up in competition with the older American cutter *Vanitie*, which had been loaned by her owner Gerard B. Lambert, who during this year and the next was to prove a staunch support of Anglo–American yachting amity at times when it was badly in need of it. It was whilst sailing with *Vanitie* that those on board *Endeavour* learned the value of the genoa jib in such large yachts, American yachtsmen had remained ahead of the British in the development of this vital sail. *Endeavour* borrowed a genoa from *Vanitie*, but during the cup races may have been handicapped rather than helped by it owing to the particular skill needed in handling such a large area of canvas.

Meanwhile *Rainbow* had been selected as the defender. Until the last moment the choice between her and the rejuvenated *Yankee* had been in doubt. Harold Vanderbilt's brilliant handling and attention to detail, and the yacht's undoubted quality in light winds, made the decision in her favour; but some people at the time and more perhaps later believed that *Yankee* would have been the better defender.

Endeavour proceeded to win the first two races for the cup in a convincing style. The crowds in the vast sight-seeing fleet off Brenton Reef appreciated that one of the few great contests for the cup was being sailed before them. And it became alarmingly evident to those responsible for keeping the cup in America that *Endeavour* was the faster yacht when there was any weight in the wind. Only in very light conditions could *Rainbow* hold her own. With five more races to be sailed if necessary *Endeavour* had but to win two of them to take the cup. Only a succession of light wind days, it might have seemed, could save the defence. Ah, but in 1920 had not *Shamrock IV* won the first two races, and the contest consisting then of five races, she had only to win *one* more for victory? *Endeavour* was half-way to winning the cup. *Shamrock IV* had been two-thirds of the way. But she lost the next three races in a row. Now, barely credibly as it might have seemed to the spectators off Brenton Reef as they waited for the third race, *Endeavour* was about to lose four in a row – and to the inferior yacht.

Despite light airs which became lighter, *Endeavour* was well out in front at the leeward mark of the third race, which she rounded more than six minutes ahead of the defender. Instead of steering straight back to the line, which seemed possible despite the vague airs, *Endeavour* was tacked several times, failed to cover *Rainbow*, and struck a dead calm, while *Rainbow*, which had not lost way through tacking, was able to creep ahead in light airs to win by three minutes twenty seconds. Bad luck coupled with unsatisfactory tactics in difficult conditions had lost the day for *Endeavour*.

The fourth race introduced the ill-feelings which thereafter dogged the contest. The clock might have been put back to the *Dunraven* and *Valkyrie* days; the beneficence of Sir Thomas Lipton have never been known. What W. H.

Taylor, writing in the *New York Herald-Tribune*, described as 'a new high in international squabbles over the racing rules' was established. It should, however, be recorded that *Rainbow* had received instructions from the Cup Committee to avoid so far as possible during the series making a protest. She therefore refrained from hoisting a protest flag when the two yachts were jockeying at the start, but it is believed – a point of importance later – that the Committee observed a breach of rules by *Endeavour* during the brisk manœuvring before the line was crossed. *Rainbow*, however, was first across. During an ensuing bout of short tacking *Endeavour* – significantly – lost further distance, a measure of the difference between *Rainbow*'s seasoned professional crew and *Endeavour*'s creditable amateurs. No less significantly, once the yachts parted to go on their separate tacks the wonderful weatherly qualities of the challenger under her quadrilateral jib, while *Rainbow* was under a genoa, carried her so finely up to windward that she rounded the next mark in the lead. After the mark, *Rainbow*'s genoa was the better sail, and *Endeavour* bore away to shift from quadrilateral jib to genoa, enabling *Rainbow* to try overtaking to windward. *Endeavour* luffed to prevent her doing so. Blue and white hulls converged until there was no more than a few feet between them. *Rainbow* failed to respond to the luff, thereby contravening a basic racing rule. Or did she? The rule itself was one difficult to apply and depended upon the point at which, had *Endeavour* maintained her luff, she would have struck the defender's hull. But since she bore away to avoid collision this point remains a matter of supposition; but general opinion both American and British believed that *Endeavour* was within her rights and *Rainbow* wrong in failing to respond to the luff. As a result of not doing so, the defender was able to sail through and thereafter was not overtaken, leading *Endeavour*, now flying a protest flag, over the finishing-line.

The Race Committee declined to hear *Endeavour*'s protest on the grounds that her protest flag had not been hoisted immediately but only as the finishing-line was approached. This was a local rule, not usual in racing elsewhere, and the Committee was entitled to enforce it, even though it was an American who wrote at the time that 'Britannia rules the waves but America waives the rules'. But had *Endeavour*, observed by the Committee, who would have been entitled to raise this protest on their own account, committed a breach of the rules at the start which would have put her technically out of the race before *Rainbow*'s foul had been committed? Since neither protest case was heard, the matter remains in doubt to this day. There is no doubt about the feelings aroused and Sopwith's outspoken comments at the time.

The race therefore went to *Rainbow*, and then the next two. The first of these is generally considered to be *Endeavour*'s one decisive defeat. *Rainbow* had ballast added between the fourth and fifth races and she was brilliantly handled by both afterguard and crew. On board the British yacht there was frank demoralization following the events of the fourth race, and in the temperamental business of first-class yacht racing the mental state of helmsmen and crew is communicated to the boat in subtle ways. *Rainbow* might, however, still have lost this race; for a man fell overboard; but the need for the yacht to be rounded up and stopped to pick him up was saved by the man's quickness in grasping a line, enabling him to be hauled back on board. Both yachts sailed round the course of the sixth race with protest flags flying owing to an event at the start. *Endeavour* had the best of the start but was tactically outwitted during the race. Both protests were withdrawn and *Rainbow* had won the race by 55 seconds and with it the contest. The yachting correspondent of the *New York Herald-Tribune* describing this race said that 'both skippers were out for blood' and added that

Yankee, the last American cutter of the Big Class to race in English waters, in 1935. She twice came close to being selected as an America's Cup defender.

there could be no doubt that if the American yacht failed to respond this time to a luff 'the British skipper would probably have been glad to cut *Rainbow* in two'. Such was the atmosphere generated by the end of the fifteenth challenge for the America's Cup.

Yet by the end of the following season Anglo–American relations were evidently repaired. The summer of 1935 brought one of the classic seasons of Big Class yacht racing in British waters (see next chapter). It was destined to be the penultimate season of the Big Class, and the last in which it appeared in strength and beauty. It might have been 1893 again, except that in some ways it was better than 1893. As in 1893 an American cutter was out racing with the British fleet; and, incredibly, as in 1893 *Britannia* was still out racing; but she was old and tired now, at last decisively outclassed. It was to be her final season.

The American cutter was *Yankee*, the yacht that twice had come close to defending the America's Cup and, in the opinion of many, the rival of *Endeavour* as the best J-class cutter afloat. Her season in English waters had as a prelude a race across the Atlantic rich in lessons which bore fruit in future yachting developments. Gerard B. Lambert, now *Yankee*'s owner, owned also the famous three-masted schooner *Atlantic*, which had won the Transatlantic race in 1905 and established a record for a crossing of the Atlantic under sail which stands today. It was planned to make of the Atlantic crossing to England a race between *Atlantic* and *Yankee*. The America's Cup challengers had been sailed and towed across the Atlantic. *Endeavour* and *Shamrock V* before her had crossed under much reduced ketch rigs, and *Endeavour*, after the cup races of 1934 had sailed most of the way back across the Atlantic having parted her tow. There was needless newspaper fuss when not unnaturally no news was heard of her for a while; but she took the Atlantic in her stride.

But she had cruised across, and her modest ketch rig was an unambitious

Atlantic, winner of the Transatlantic race in 1905 in a time of 12 days, 4 hours and 1 minute, a record that still stands. This famous schooner accompanied *Yankee* (see p. 179) to England thirty years later, the two yachts racing across the Atlantic. Victory went to the much smaller and more modern cutter.

cruising spread of canvas. *Yankee* had her racing mainmast cut down only to the jib halyards; she was therefore able to set ahead of the mast her full racing canvas. A short mizzen mast was stepped aft to convert her into a yawl, but so small was the sail set on it that she would probably have been better without the additional complication; essentially *Yankee* emerged from her ocean-going treatment as a handsome ocean racing cutter. She was 126 ft long overall and 86 ft on the waterline; *Atlantic* 185 ft and 137 ft. The latter was representative of the older concept of ocean-going yacht of large size; the former the yet last word in evolution of the racing yacht. *Atlantic*, with her much greater length, had the advantage of a considerably higher maximum speed; *Yankee* had the class racing yacht's close windedness; she could point high on the wind, which was *Atlantic*'s weak point of sailing.

The two yachts sailed much the same course until the final stage of the race, and *Yankee* took the lead and held it the whole way, though they were within sight of one another on the twelfth day out and on the fourteenth they were almost level, though with *Atlantic* to the south and less well placed for the finish off the Lizard. *Yankee* finished some seventy miles ahead; winds from before the beam undid *Atlantic* in the later stages.

Two such disparate yachts might each win a race such as this given the particular conditions to suit their abilities. The important revelation of the race, however, was the comparative behaviours of the two yachts in heavy weather. On the ninth day out there was a gale estimated by *Yankee*'s experienced navigator Alfred Loomis as Beaufort Force 9 gusting 10. *Atlantic* laboured heavily, and with two men at the wheel was reduced to storm trysail, double reefed mainsail and two headsails, and later to trysail and single headsail. Sailing thus a sea carried away twelve feet of the bulwarks, after which the ship was hove-to.

Yankee was never hove-to and under the worst of the conditions Loomis described her action: '*Yankee* behaving beautifully . . . our motion below very quiet and we are logging 12 knots under staysail and trysail.' On this day *Yankee* made 273 miles compared with *Atlantic*'s 296 miles. It was a beautiful demonstration of the inherent ability of a twentieth-century racing machine; the hulls of the J-boats, in their basic design and light construction of great strength had provided a new standard of speed and seaworthiness, and their excellence was only marred by their fragile racing rigs. As the author wrote in *British Ocean Racing*: 'The relative performance of . . . *Atlantic* and *Yankee* in 1935 throw into high relief the progress of yacht architecture that would not be easily detected otherwise; for close studies of the past (represented by *Atlantic*) in relation to the present are rarely made and in the case of sailing ships' performance are always difficult owing to the sparseness of information.' *Yankee*'s race across the Atlantic underscored the facts of design for ocean-going that were being learned in the offshore racing world.

Yankee's season in English waters is considered in the next chapter.

11
The international sport

While the Big Class sailed, and sometimes faltered, through the seasons, a hitherto unprecedented interest had sprung up in the smaller International rule classes, the 12-, 8- and 6-Metres. It was curiosity in inter-war regatta yacht racing that a wide gap of size existed between the biggest class and the next largest, the 12-Metres. The J-boats were to reach the great size of some 130 ft overall, 87 ft on the waterline, and carry a displacement tonnage of about 160 tons. The next largest class, the 12-Metres, though big enough to dominate any regatta by their size and thoroughbred elegance after World War II, appeared small indeed beside a J-boat, with nearly double the length, more than four times the sail area and about six times the displacement tonnage.

This was the gap that prior to 1914 had been filled, not very adequately, by the 15- and 19-Metres. Now it yawned, appearing illogical, though such are the chances and whims that govern yacht-racing taste that it is impossible to say

whether an intermediate class would have gained a following. The matter was much discussed at the time, and in 1936 steps were being taken by leading yachtsmen to establish such a class. It is significant that amongst them were a number of owners in the J-class; suggesting that the charms of great size were thinning under the friction of manning and maintaining such obvious sailing grandeur. But while the big craft were then just about to disappear for ever, no intermediate class appeared. Had it done so the war on the threshold would have sent it the same way into extinction.

The 6-Metres were one of the greatest classes in the history of yacht racing. Between 1920 and 1939 they were the smallest of the International classes; in technical development they were second to none. By the 1930s they had evolved into boats with the following approximate dimensions: length overall 38 ft, length waterline 23 ft 6 in, beam 6 ft 0 in, draught 5 ft 5 in. The rated sail area might have been about 450 sq ft, an occult figure bearing a certain relationship to the actual amount of sail sometimes carried, which latterly included a parachute spinnaker which alone contained more than double this area of fine, fragile cloth.

It will be evident that the 6-Metres were very narrow for their length, and also rather heavy considering that they had no accommodation. They would have been narrower still but for the fact that the rule makers had stepped in when some boats had appeared with 5 ft 9 in of beam, decreeing a little belatedly that in future beam should not be less than 6 ft. Since the Sixes had a displacement

Below left

The 15-Metres out racing in 1912. This size of yacht was missing from the European racing in the period 1920–39, leaving a considerable gap between the 12-Metres (see p. 195) and the Big Class to the Universal J-class rule.

Below

M-class yachts to the American Universal rule racing in 1935, *Prestige* leading. Yachts of this size were intermediate between the 12-Metres and the J-class, and might have served this purpose in Europe during the inter-war years.

tonnage suitable for a boat with some accommodation and interior fittings, but had none, the weight that might have gone into them went into the keel, which in later boats totalled more than seventy per cent of the total.

For essentially open boats with nothing in them except the best facilities possible for handling what had become by the 1930s a most expansive equipment of sails, the Sixes were excessively expensive pure racing machines. Legislation following World War II killed the class, though it was too good to die quickly, while too extravagant to live much longer.

In 1920 there were no 6-Metres in the U.S.A., and to most yachtsmen and their architects the European international rule of measurement was as little considered as was the American Universal rule in Britain. By the season of 1924 there were some fifteen American Sixes built between 1921 and that date, and handling them was an aspiring group of mainly young racing men occupied in establishing new standards of sailing expertise.

Some doubts surround the origin of the British–American Cup, first raced for by the two national teams of Sixes in 1921, and becoming, along with the older Seawanhaka trophy, the source of the most dedicated international racing of the era, less grandly spectacular than the America's Cup but richer in the fine art of race sailing. The story of the British–American Cup is one of an initial British ascendancy eroded all too quickly as the Americans learned about a type of boat hitherto strange to them, and evolved techniques of handling hitherto strange to everyone. The idea of the cup for team racing appears to have originated when *Shamrock IV* was racing for the America's Cup in 1920. Arrangements were

American 6-Metre *Nancy* in the season of 1932. *Nancy*, when racing in the Solent, revealed to British owners in the class the value of a big wardrobe of headsails, at a time when British Sixes might have carried two only.

Right

The International class next larger to the 6-Metres, a group of 8-Metres is seen here racing in the Solent. Typical dimensions of an 8-Metre were length overall 49 ft, length waterline 30 ft 6 in, beam 8 ft 3 in, draught 6 ft 6 in, displacement 10 tons, rated sail area 830 sq ft.

Below

Seawanhaka cup.

made for the contests to be held annually, alternately in England and America.

In 1921 the American team was decisively beaten at Cowes, and again, though after closer racing, at Oyster Bay in the following year. The story was repeated in 1923, when a team of light-weather American Sixes was defeated in heavy Cowes weather. By 1924 the British had been designing and racing 6-Metres for thirteen seasons and had built some forty boats before the Americans possessed a handful. In 1924 the British team won again, by a narrower margin this time, and in accordance with the terms of the races the cup passed permanently into their hands. A new cup was subscribed for jointly, races being established now as biennial affairs, and competition was resumed in 1928. This cup was won outright by the Americans; and so too was a third.

In 1922 the old Seawanhaka cup was put back into circulation, after lying neglected in the Manchester Yacht Club, Massachusetts, since early in the century, the racing of scows having lost its appeal. It was now presented for match sailing in the 6-Metre class, and this brought a new and greater international fame to the trophy, which became regarded as the America's Cup of small yacht racing. In 1922 the Scottish 6-Metre *Coila III*, the champion of her day, easily defeated the indifferent American Six *Sakie* in three straight races; but the fact that the cup had been carried to Europe assured its future. In 1923 it was defended on the Clyde by *Coila III*, which again won, but this time by three races to two. In 1925 the cup, its revived reputation assured, returned to the U.S.A., when with *Coila* still the defender, the American Six *Lanai* was victor in three straight races. *Lanai* was designed by Clinton Crane, who was later to produce the J-class *Weetamoe* for the defence of the America's Cup, while at the helm of *Lanai* was Sherman Hoyt, one of the century's most brilliant performers in all types of yacht racing. By the mid-1920s it was clear that the 6-Metres were sailing out into the van of racing development; in design, rig and sail handling the superb little craft and their skilful crews were teaching lessons to the larger 8- and 12-Metre classes and the J-boats.

An unsuccessful experiment, the *Atrocia* of 1927, provided a measure of the experimental attitudes prevalent in the class. With her mast stepped abaft of amidships, and a large jib with correspondingly small mainsail, she was feeling the way towards the overlapping genoa jibs that were about to transform the sail plans of all racing yachts. But in *Atrocia* there was no overlap to the jib, which carried a boom along its foot longer than the main boom. She was not a success; nor was a 6-Metre similar in general conception which came out in Britain; but they were brave anticipations of the future now just on the doorstep.

In the Sixes there first appeared at this time three developments in sail design and handling not less revolutionary than were the cotton sails of the *America* in comparison with the flax-sails of Solent yachts of 1851.

Since it first appeared in its modern form in 1927, the sail known as a Genoa jib – now familiarly contracted simply to 'genoa' – has become the seat of power in sail plans, while mainsails, no longer logically so named, have become guide vanes to lead off the wind from the dominant genoa. Overlapping jibs and headsails were an ancient seafaring device in 1927, made of lighter cloth than the working sails and used in reaching winds and light airs. The 'mumblebee' or reaching foresail of the Brixham trawlers used to be a traditional example of such a sail, of which it was said 'anyone who has ever seen the monster foresail of a Brixham trawler set and drawing can well understand the wonderful pulling power of a balloon foresail when reaching with the wind anywhere near the beam or abaft it'.

So when a sail apparently resembling this was carried in a 6-Metre during the Genoa regatta of 1927 history might not have been made. But when racing later in the year in Oyster Bay, the way in which the sail was set and handled produced the revolution. This was due to the enterprise of Sven Salen, a leading Swedish helmsman, who handled the 6-Metre *Lilian* in Genoa and *Maybe* in the U.S.A. As the yachts rounded a leeward mark to turn on to the wind, the big overlapping jib was not, as hitherto would have been conventional, lowered and replaced by a smaller, flatter sail. Instead the big jib was simply hardened in, sheeted taut for windward work; and it was demonstrated that rigged thus the boats clawed up to weather fast. The reaching ballooner of light canvas had been converted into a heavier, flatter-cut headsail capable of being harshly trimmed for beating. It was a new technical concept destined to move quickly from the 6-Metres to all types of fast yacht.

It was Salen also, and again in a 6-Metre, who appears to have taken the first steps in evolving the parachute spinnaker, which has since dominated yacht racing and become over-loved by photographers. The spinnaker was initially a variant of the square sail suitable for amalgamation into the fore and aft rig when running off the wind. The old style of spinnaker was a relatively flat sail sheeted inside the forestay and becoming crushed against it under some conditions of trimming. Such a sail was not unlike a ballooner, triangular and having a luff and leech. The new spinnakers were double luffed and symmetrical about the middle line and cut not flat but with increasing amounts of balloon in their shape until they became like circus tents, big enough to wrap up the boat, and exceeding by twice or more the total area of the working fore-and-aft sail including genoa.

Great straining, brutally sheeted genoas pinned in as hard as sweating men and geared winch could contrive; gossamer spinnakers whose fairy bubble forms belied the immense loads that they put upon mast, sheet and guy at their three points of attachment – these developments of rig given to the world of yachting by the 6-Metres transformed the sailing scene into the essentially modern one known today. Yet these inventions are rich in the irony of the earnestly pursued illogical. They were produced in a day that was busy eschewing the sail clouds of the richer past, the mainbooms swinging far beyond the counter, the bowsprits reaching out ahead into the middle distance, and those crowning glories of topsails scratching skies with their jack and topsail yards. These splendours had been replaced by the restrained Bermudian rig, short-boomed and chastely bowspritless, tall, cool and economical. But to carry sail to the limit of the possible is the unquenchable urge of the racing seaman. Like sin,

Bobcat, a well remembered American 6-Metre.

curbed in one direction, it bulges in another.

The American climb to mastery in 6-Metre racing, which had become the most exacting test of any in sail handling and design, was most convincingly demonstrated in the Solent in 1932. It has been seen above that the first series of British–American Cup races had ended with the trophy being won outright by the British after four successive annual victories. But now the Americans were on their way to winning outright the second trophy, after victories in 1928 and 1930.

In 1932 their team won four straight races from the British, making it unnecessary for the remaining three of the series to be sailed. This potent demonstration of superiority was partly due to the efficient techniques of team sailing that had been evolved by the Americans, while the British team of four boats tended to race as individuals. But – the long shadow of the *America* reaching down the seasons – it was American sails and cunning American handling of them which assured their victory. The parachute spinnaker was yet a novel sail; techniques for handling it were still being evolved. Before the invention of the parachute spinnaker, the sail had been used exclusively for sailing downwind. Now it was found that they might be kept set while the wind came on the beam; set very shy indeed they might be carried for brief periods with the wind edging ahead of the beam. This was a revolutionary new art in sail handling.

The genoa had become established, and it led to a transformation in the sail lockers of racing craft, beginning in the Sixes. The boats in the British team in 1932 carried only two headsails, apart from the spinnaker, a large genoa having the maximum overlap allowed by the rule, and a smaller jib with its tack about level with the mast. Thus was *Vorsa*, probably the best boat in the British team and designed by that old master, Alfred Mylne, who had produced some score of 6-Metres before any American architect had produced one. In the American team was *Nancy*, designed and owned by the young Olin Stephens, and so far as the basic hull was concerned perhaps no better than *Vorsa*. But while the latter carried the mere two headsails, *Nancy* carried five stuffed into the bulging sail locker under the foredeck. The genoa and the smallest sail but one matched the two of *Vorsa*. There were two intermediate jibs between these, sails to which the Americans owed much in the 1932 races, when the exiguous British sails lockers were unable to provide fine graduations to suit the wind; and finally *Nancy* had a very small jib for the worst conditions. Such an outfit of headsails, now commonplace in first-class racing, was like the genoa and the parachute spinnaker at that time, a surprising advance in technique, the third to emerge from the 6-Metre class.

Endeavour and *Yankee* were the stars of the 1935 season. We have described above *Yankee*'s passage across the Atlantic. When she arrived in England it was Jubilee year. In July a naval review assembled in Spithead, and in two of the columns flanking those of the major warships, amongst the liners and channel steamers, some yachts had been assigned anchorages. During the few days of the assembly others of all sizes passed between the grey lines of the warships. '. . . we were not prepared for the sight which greeted us when we completed the crossing and anchored off Spithead, close to a group of naval vessels.' Thus wrote the owner of the *Yankee*.

The weather was rich in English contrasts that year, blowing hard at the beginning and end of the season, with light winds between. There were sometimes eight yachts racing in the Big Class, rarely less than six: *Endeavour*,

Yankee and *Velsheda*, the three J-boats rating at the 87 ft maximum of the Universal rule raced level; *Shamrock V*, to the same rule but of slightly lower rating, received time allowance, as in increasing quantities did *Candida*, *Britannia* and *Astra*, the last being the baby of the fleet. The schooner *Westward* was out for eleven of the races, dominating the scene by sail-crowded magnificence wherever she appeared, larger and faster than the J-boats when strong reaching winds gave her the conditions that best suited her; and they brought her two first prizes. *Astra* was second only to *Endeavour* in the number of flags she gained that summer, both yachts sailing in thirty-five races. *Yankee* sailed in thirty-two. In the first race in which *Endeavour* and *Yankee* competed together *Endeavour* was dismasted, and *Yankee* was dismasted in the last; by which time *Endeavour* had beaten *Yankee* in nine races and *Yankee* beaten *Endeavour* in eight. It was a satisfactory score between what many regarded as the two best J-boats in existence. *Britannia* was not to win a single race in 1935, or a flag for second or third place. At last the curtain had dropped on the longest career of any first-class racing cutter. She could no longer revel in even the hard winds which, for so many years, had been her particular joy.

That year of 1935 saw the climacteric of British Big Class racing. But very few realized it; perhaps only Charles Nicholson, the man who had so largely created the fleet that had been out that season – five out of the eight boats were the productions of his pencil and building yard, and many more in the 8- and 6-Metre classes. He spoke then of the sunset while others were dazzled at what seemed a new moon of affluent yachting. The decline of the Big Class, despite the appearance of a new yacht for T. O. M. Sopwith – *Endeavour II* – was revealing itself in 1936. In April of the year *Yankee* sailed home to the U.S.A. *Shamrock V* was not fitted out; nor was *Candida*. *Westward* did not race. *Britannia* was given a grave at sea, and as she sank the Channel waters closed over more than forty years of yachting history.

There were only four boats in the Big Class of 1936, *Endeavour II*, *Endeavour*, *Velsheda* and *Astra*. The new *Endeavour* was expected to be a challenger for the America's Cup, but no challenge had yet been issued. Twice in the course of the season she was dismasted, and this frequent loss of masts in the Big Class – boats which, apart from their rigs, were of such well proved excellence of design and construction – was causing widespread unfavourable comment; never so strong as when, on days that should have brought inspiring racing, the Big Class remained at their moorings for fear that a capful of wind would collapse their high rigs. Persistent criticisms encouraged one of the rare contributions to the press by Charles Nicholson. Appearing in the *Yachting World* (London) on 21 August 1936 it has considerable historical interest both because of its contents and the rarity of his public pronouncements:

'The truth as to the causes of recent carrying away of masts of large racing cutters should be known, because much has been said and written which is incorrect – putting it very mildly – about their recent "failures", which many yachtsmen, and others interested, too readily swallow.

'The habit of inaccuracy is possibly acquired by Press urgency to send immediate copy before writers have taken time to enquire the truth, and many will not take the trouble to do so from a proper source.

'The facts are as follows:

'First, in reference to the causes of the last casualties with the gaff rig, as a reminder to those who live in the past and still prefer and praise that ancient rig, forgetting it was not immune from these troubles.

'In 1927, *Lulworth*, gaff rigged, carried away her solid mast in her last race through the breaking of a lower shroud bottle screw.

'Later, *Shamrock* 23-Metres, also carried away her solid mast through the same cause.

'In 1928, *Astra*, in her first season, lost her solid mast. Its height above deck was restricted by Y.R.A. to 127 ft compared with 155 ft of the hollow mast she had in 1931 on our adopting the N.Y.Y.C. rating for classes above 12-Metres, and she has since carried it without trouble. On trial I found she put undue strain on the forward lower shroud in fresh winds, presumably owing to the great breadth of the lower part of mainsail due to the long boom, which the short mast made necessary. This bent the lower part of the mast aft. To check this, a portable inner fore stay was fitted for use in fresh winds. In the first hard blow, with double-reefed mainsail, this stay was not used. The excessive strain on the forward shroud broke it and away went the mast.

'*Cambria* very nearly lost her solid mast on the same day from the same cause. The upper splice began to draw, and on hearing it Sir William Burton saved the spar by risking a quick luff.

'The next casualty in the J-class, five years later, was with *Shamrock V* in 1933, racing at Babbacombe in a hard wind. Rounding a mark in a very heavy squall and bearing up with mainsheet flat, heeled more than 45 degrees, a lower shroud bottle screw parted. I should add that although it had survived four seasons, it was found, on testing one of the sister screws, the metal was not up to its normal strength.'

Ranger. Notice the quadrilateral or double-clewed jib and the diagonal-cut or mitred mainsail.

During Cowes Week in 1936 steps were taken which, had war not intervened, might have radically transformed first-class yachting. It is possible to believe that without the intervention of war the J-class would still have been doomed, not because of their fragile rigs, which could have been remedied by suitable rules, but by their expense and the problems of manning such large craft. They were becoming anachronistic in the social conditions of the late 1930s.

The strangely large gap in size that existed between the J-class and the next largest class of 12-Metres has been noted above. At Cowes in 1936 the council of the Yacht Racing Association set in motion the machinery required to bring into being another class of intermediate size. The American Universal rule governing the J-class was selected, the size to be 56 ft rating, compared with the 76 ft rating of the J-boats. The yachts would have been comparable with the old 19-Metre class of pre-1914, and as an authority said at the time, they would have been 'quite important enough to form our largest class'.

Prior to this meeting in Cowes, C. R. Fairey, then the owner of *Shamrock V*, had issued a challenge for the America's Cup to be sailed in yachts of this intermediate size. The challenge was not accepted. Subsequently T. O. M. Sopwith issued his second challenge, which was accepted. The terms were changed in a crucial respect from those formerly governing the contests. The challenger was not named. By the end of the 1936 season the first *Endeavour* had won seven races out of thirty-two sailed; *Endeavour II*, joining the fleet late, sailed only eighteen races and won nine. But there was some doubt about the relative merits of the two boats, especially as the older of them had not appeared to be sailing up to her best form that year. When Sopwith issued his challenge, to be contested in 1937, it was mutually agreed by the two sides that either of the yachts might be the challenger, the one selected not having to be named until a month before the contest.

We have considered above the contest of 1934 in some detail. The contest itself in 1937 does not merit such attention for the races were a sad procession with the challenger trailing far astern of the defender in four races. The defender, however, whose active life was so short, rising to the brief climax of the four Cup races, was a yacht of such brilliant performance that her place in history is assured. With *Endeavour II* Charles Nicholson produced a yacht that was essentially an enlargement of the first *Endeavour*; for it had now become clear that the fastest J-class yacht must be one designed to the maximum waterline length allowed under the rule. The defender, however, was a revolutionary concept, one of the small band of racing yachts that proves immediately to be far in advance of what has been before. The manner of her creation anticipated the process of design that was to become dominant henceforward.

The boat that emerged, *Ranger*, was one of the rare America's Cup defenders not built for a syndicate. Harold Vanderbilt undertook the defence alone after difficulty had been encountered in raising funds. The necessity for a new defender was unusually pressing in 1937. *Endeavour II* was probably, it was felt, faster than *Endeavour*, and the latter was at least a match or even faster than *Rainbow* and *Yankee*. Vanderbilt commissioned a design to be worked out jointly by Starling Burgess, designer of the last two defenders, and Olin Stephens of Sparkman and Stephens, then at the beginning of a career that was to rival that of the great Nathaniel Herreshoff himself. Burgess later wrote an account of the creation of design of *Ranger* in the journal *Yachting* (New York):

'Vanderbilt ordered the design of *Ranger* from Sparkman and Stephens and myself last fall. *Endeavour II* was already built, to the 87-foot upper limit of the class and had made a creditable showing against *Endeavour I*. Olin Stephens and I found at our disposal the new towing tank of the Stevens Institute of Technology in which Professor Kenneth S. M. Davidson had worked out a most extraordinarily ingenious method of testing sailing models in which the useful driving force of the wind, its heeling force, and its component of leeway were closely resolved by the mechanical resistance of springs and balance weights. For testing a model's ability to go to windward, we were enabled to plot speed made good to windward directly against wind speed.

'Nicholson had given me the lines of *Endeavour I*, so our first step was to try out, as measuring sticks for the new model, the models of *Endeavour I*, *Rainbow* and *Weetamoe*. We were encouraged to find that the models of these three boats gave results strictly in accordance with the observed performance of the full-sized vessels.

'Vanderbilt, Stephens and myself were fully agreed that we must go to the 87-foot hull. Four parent models were constructed, and departures in the shape of overhangs, position of the rudder post and so forth, were tested in most of the parent forms. The model selected for *Ranger* was so unusual that I do not think any one of us would have dared to pick her had we not had the tank results and Kenneth Davidson's analysis to back her.

'However, not only did the dial readings indicate her as the best of the lot, but her photographs showed a wave formation much smoother than that of the others.'

There was a common belief that the wonderful hull form of *Ranger* had been the work of the young Olin Stephens, a belief that was repeated in Burgess's obituary in a leading English Journal. It was perhaps natural for this to be expected; not least in the light of Burgess's near failure with *Rainbow* in 1934. The vision of the older and more experienced architect being at hand to steady young enthusiasm, while no longer capable of the higher flights of creativeness able to produce so outstanding a hull, is too obvious not to have appealed and gained wide credence.

In fact, as Olin Stephens has now explained, the collaboration between the designers was based on an agreement that all phases of the work were to be a common responsibility, whoever performed it, and that the latter would be confidential. Even Harold Vanderbilt retained the opinion for more than twenty years that Olin Stephens had drawn *Ranger*'s hull lines, which was then denied by Olin Stephens himself. The false attribution was repeated by John Illingworth in his *Twenty Challengers for the America's Cup* in 1968, and Stephens occupied most of the Foreword he wrote for the book, again denying the attribution of the design.

Sets of hull lines for *Ranger* were drawn by both Burgess and Stephens, and it was from lines by the elder designer that the model ultimately selected after the series of tank tests was made. Thus Burgess may be said to have redeemed, and outstandingly so, a suspicion that in the matter of producing hull forms he had met his master in Charles E. Nicholson. *Enterprise*, as we have seen, was possibly a less effective form than *Shamrock V*, and *Rainbow* compared with *Endeavour* certainly was; but Burgess had clearly come to the top with *Ranger*.

The attribution of all the other considerable work involved in producing *Ranger* – work shared by Olin Stephen's brother Rod, now internationally recognized for the brilliance of his talent for all rigging matters – remains confidential, and we may regard the yacht as the outcome of a successful partnership.

The *modus operandi* followed by the team of Burgess, Stephens and Davidson in producing the new design appears to have been as follows: As Burgess mentions above, the models of *Endeavour* (Nicholson), *Weetamoe* (Crane) and *Rainbow* (Burgess) served as the essential yardsticks. It will be noticed that no model of *Enterprise* was tested, for by this time it was evident that dimensionally with her short waterline she was too far wide of the mark to be of comparative interest. Burgess remarks above that it was generally agreed that the new design must be of 87 ft waterline, the maximum allowed by the rule and selected a year earlier by Charles Nicholson when he produced the drawings for *Endeavour II*; and it is of interest to appreciate that this crucial and correct decision by Nicholson was not the result of tank tests.

Burgess and Stephens then produced several alternative designs working independently, and from the best of them, as assessed in the tank, a parent form was produced and subjected to further modifications in detail. Thus the surprising form of *Ranger* emerged, which could not have been achieved without Burgess's flair in conceiving the initial form, and the tank testing technique which enabled it to be refined and proved acceptable.

The most striking feature of the hull which distinguished *Ranger* stood out from any other boat of her class and was of relatively slight importance; the rounding of the deck in plan at the stem head and in profile the rather graceless blunting of the forward overhang which, in other J-boats, and most exaggeratedly in *Endeavour II*, reached out to the limit of the soft, fair curve of the profile, to terminate the hull in a sharp point little more than a foot ahead of the outer forestay.

A model being tested in the Stevens Tank, Hoboken, U.S.A.

Ranger proved so decisively superior to *Yankee* and *Rainbow* that though little was yet known about the performance of *Endeavour II* the betting odds quoted before the contest were two-to-one that *Ranger* would win by taking four straight races in a row. And this is what she did. She won the first race by seventeen minutes and the second by more than eighteen, inflicting on the lovelier blue yacht two of the biggest defeats in America's Cup history. These were light wind races of the kind that Sopwith had hoped to avoid by sailing the contest earlier in the season than usual. With more wind in the third race *Endeavour II*'s form improved, but she was defeated by four minutes and by a little less than this in the fourth. It is undoubted that the challenger would have made a better showing in harder winds, but whatever the conditions *Ranger* must have proved the victor. When Charles Nicholson saw the yacht out of the water he pronounced her hull form the most revolutionary advance in design for fifty years.

The races over, in a hush of defeat that was to be repeated in the British America's Cup challenges of 1958 and 1964, no reason for it was more strongly impressed on the public mind than the lack of tank testing in Britain; and in 1946, when the return was being made to yachting though no America's Cup racing was in prospect, Charles Nicholson said 'It is quite useless for us to hope to compete successfully in Anglo–American class yacht racing before our designs can be properly tank tested.'

Immediately after the 1937 contest, Major B. Heckstall-Smith wrote shrewdly on this subject:

'So far as tank tests are to be considered, it should not be forgotten that they are something very nearly new to British ideas on designing sailing yachts. Yet they will confirm that one hull is better or worse than another hull undergoing the same tests.

'Tank tests decided the choice of *Ranger*'s hull and that means that the Americans were just that much smarter than those responsible for *Endeavour*. The Americans checked up that bit more but it does not by any means follow that the same sort of tank tests would have produced a better British hope. They might merely have confirmed that *Endeavour II* was the best we could produce.

'It is certain that small blame can be attached to Mr Sopwith or his colleagues, easy though it is now for the armchair critics to blame him for the lack of tank tests. No one before the defeat of *Endeavour II* seems to have had the idea of suggesting that tank tests should have been carried out on this particular British boat.

'Now that the tank tests and data have been disclosed, however, this form of research is almost certain to play an important part in the future of yacht designing, which is, assuredly, a step forward in the right direction, and certainly proved to be a safer touchstone than judgment on lines only, or the application of some theory of balance.'

After the cup races in 1937 there was the New York Yacht Club's annual cruise, when for the first and last time five great J-class sloops were out racing together, and for the last time there was racing between the biggest type of dedicated racing cutters of the breed originated by Nathaniel Herreshoff in the 1890s, of which the J-boats were the final representatives.

Of those five boats, superficially very alike but for the dark colours of the English boats and the blunt aggressive stem of *Ranger*, one was by Burgess alone – *Rainbow* – with *Ranger* by Burgess and the Stephens, two were by Nicholson, the *Endeavour*'s, and *Yankee* was by Paine, whose father had sponsored the father of Starling Burgess. The three other J-boats designed in this last era of big yacht racing were absent – the first of the J-class, which had come out in 1930 – *Shamrock V* and *Enterprise* by Nicholson and Burgess, and *Weetamoe* by Clinton Crane.

The winter of 1937 came down on a kind of sea beauty that man was to build no more.

There was no Big Class in Britain in 1938, and there was never again to be one; the long laid-up mud-berth days had begun for the British J-boats. The 12-Metres became the largest European class racing yachts, now with several former

J-class owners sailing them. In 1939 T. O. M. Sopwith had built the 12-Metre *Tomahawk*, and in that year he met again his rival Harold Vanderbilt, also now racing a Twelve.

He brought the new 12-Metre *Vim* over for a season of British racing in 1939; and with this yacht appeared the shape of things to come; though they were over the far side of six years' total war. In 1939 nobody could have expected that nineteen years hence the America's Cup would be raced in boats of the relatively small 12-Metre class; nor that *Vim*, twenty years old by this time, would be good enough to be almost selected as the cup's defender. But the future was

The American 12-Metre *Vim* racing in the Solent in 1939. Designed by Olin Stephens and sailed by Harold Vanderbilt, she was outstandingly superior to any European or American Twelve, representing an advance in basic design and equipment which was to be of great benefit to the U.S.A. when America's Cup racing was resumed in this class in 1958.

latent in *Vim*'s performance in that last season of European yachting before the bombs.

Brooke Heckstall-Smith, who forty-six years earlier had watched the performance of another much larger American cutter – the *Navahoe* – racing in British waters during the great season of 1893, wrote at the end of the 1939 season, when *Vim* was still in the process of being shipped back to the U.S.A.:

'*Vim*, designed by Olin Stephens and built by Henry B. Nevins, may be justly described as the most remarkable and successful American cutter that has ever visited our shore. . . . *Vim* lost her first race at Harwich through going the wrong side of a mark. She sailed seven races in the shallow East Coast tidal waters and won five. She then proceeded to the deep waters "Down West", where she again sailed seven races and won six. Then she came to the tidal waters of the Solent, where she sailed fourteen races and won eight. When we consider that all these races were sailed against our most experienced yachtsmen, in strange waters and over unknown courses, and without a pilot, and that, on the whole, *Vim* lost nothing but very often gained by superior pilotage or choice of the least tidal water, our praise for the ability of her crew cannot be too high.'

Above

Olin Stephens (left) examining a tank testing model of the 12-Metre *Intrepid*. The fin-keel and skeg of the yacht is to be noted, a late development in designs to this class.

Opposite

Large models of *Ranger* and *Endeavour II* in the New York Yacht Club's model room, where similar models of all other America's Cup challengers and defenders are displayed.

Her crew included three Scandinavian paid hands with the skill of their breed, so often proved in international yachting. The expert amateurs on board included the owner, Harold Vanderbilt, who had sailed to victory the last three America's Cup defenders, on one occasion, as we have seen above, when the dice were loaded against him, and Rod Stephens, brother of the designer, whose most active years of rigging and tuning-up of racing yachts were yet to come after 1954. The combination of talent and expertly applied brawn on board *Vim* was no doubt superior to anything found in the British Twelves. The effective pilotage mentioned above was, incidentally, partly due to the use of a dipstick, marked to show a few inches more depth than the draught of *Vim*, which enabled her to work closer inshore than yachts using the less precise judgement of pilots with local knowledge.

Throughout the 1930s yachting in the Solent reached annually fresh peaks of activity. The demand for paid hands, in these their last days of respectable numerical strength, exceeded the supply; for so many of the coming generation did not take to that way of life, and at last the Solent, after so many centuries of richness in them, was being drained of professional sailing seamen. The builders were busy on the many types of large yacht, sail and power, that the famous Solent yards still found to be the only profitable work – yachts each individually built of prime timber to the sounds of adze and saw and caulking mallet amid the smells of newly cut pine and fresh varnish.

The senior yacht clubs of the Solent were fine old brandy by this time. The Royal Southampton Yacht Club in the middle of High Street, Above Bar, was to remain standing, like the ancient Bar Gate itself, when so soon now most of High Street was going to be crashing around it. From Ryde the Royal Victoria Yacht Club, ninety-five years old in 1939, looked across towards the Spithead forts, for which there were yet no available anti-aircraft guns; and at the other end of the Island the Royal Solent, sixty-one years old in 1939, looked from Yarmouth towards the Jack-in-the-Basket off Lymington. The 124-year-old Royal Yacht Squadron waited for another war in its 450-year-old castle. Near the Royal Pier in Southampton the formal and satisfying building of the Royal Southern Yacht Club carried, in 1939, its 102 years with dignity; and very soon people standing with their backs to the clubhouse were going to be watching the German liner *Bremen* moving slowly down Southampton Water as the first bombs were falling on Warsaw.

12 Off soundings

Racing over oceans in well manned well fitted yachts was a far from unknown activity before World War I. But it was a rich, relatively rare pastime, and a little eccentric. Professional seamen and a few enthusiastic amateurs, the latter sometimes with seagoing ability, formed the crews. So far as the amateurs were concerned, the publicity usually exceeded the seamanship.

As late as 1966 a book entitled *The Great Yacht Race* displayed the following blurb:

'In December 1866 three American millionaires raced their 200-ton yachts across the Atlantic from New York to Cowes. The stakes totalled 90,000 dollars and the winner made the crossing in under fourteen days – a very little more than the fastest time then achieved by a steamship. These frail schooners were designed for sailing in inland waters, but were driven through the full fury of an Atlantic storm. Six men were washed overboard, yet skill and courage never faltered; the little ships sailed on to complete the first ocean race in yachting history.'

The above suggests that the public attitude to sailing in yachts over oceans in 1966 was hardly more sophisticated than in 1866. This famous contest between the schooners *Fleetwing*, *Vesta* and *Henrietta*, all of a little over 200 tons, has settled into the niche of yachting sagas, partly because the race was sailed in mid-winter, more definitely because the *Henrietta* was owned by J. Gordon Bennett, one of the more flamboyant, and entertaining, amongst the crowd of big spenders which was a feature of American life at the time. Involved in the organization of the race was Leonard Jerome, whose daughter Jennie became the mother of Winston Churchill.

There were four transatlantic races for large yachts held between 1866 and 1905. The fastest of the passages was that of the three-masted schooner *Atlantic* in the 1905 Spanish race. Her performance against a modern yacht, much smaller in size, is considered in Chapter 10.

In August 1921 the gaff-rigged ketch *Typhoon* was lying in Cowes after a transatlantic crossing of twenty-two days. The little dark hulled, short-masted ketch showing the stars and stripes from a disproportionately heavy counter stern, which seemed out of harmony with the fine, flared bow, did not exhibit any high degree of Solent polish in her appearance, which was homely though seamanlike. But she had aroused much interest in the port. With a length overall of 45 ft, a draught of 6 ft and displacement of 15–16 tons she was considered a small boat for an Atlantic crossing in 1921; and while the personality of her owner charmed all who met him, the shape of his boat – the motor-boat like stern and lean hungry bow – did not satisfy many who were qualified to judge.

The owner was William W. Nutting, Editor of the New York journal *Motor Boat*, who had sailed with a crew of two friends to see the Harmsworth power boat races. He stood yet on the threshold of the greater reputation he was to

The first *Endeavour*, painted by Frank H. Mason.

William W. Nutting, who inspired the foundation of the Cruising Club of America and became its first Commodore in 1922.

gain in the world of yachting. While in Cowes he became a friend of Claud Worth, who had described to him the objects and methods of the Royal Cruising Club, of which he was a prominent member. This proved to be a seed time, with the harvest soon to follow. *Typhoon*, on her tougher east to west voyage back across the Atlantic, carried on board the seed of the Cruising Club of America; and also, as it happens, amongst the crew of four this time, a young Cowes man, Uffa Fox, who like Nutting was to become one of the more colourful yachting figures of his generation.

Various happily informal meetings were held in New York during the ensuing winter between groups of Nutting's friends, culminating in one on 15 May 1922 when a body was founded and given the name of Cruising Club of America, described at the time by Nutting as 'a sort of American equivalent of the Royal Cruising Club'. There were thirty-six founder members, who included the Boston yacht architect John Alden, Frederic Fenger, Herbert L. Stone, and two honorary members, W. P. Stephens and Thomas Fleming Day – thirty-six men who, as John Parkinson, historian of the Cruising Club of America, has said 'will ever be remembered as seagoing gentlemen'.

The rules, as published in the first club handbook in the next year, closely followed those of the Royal Cruising Club. Initially the Cruising Club of America's objects were precisely described by its title; but human organizations have all the eccentricity of human personality. When, a few years later, ocean and offshore racing was to take root in Britain, mainly thanks to certain members of the Royal Cruising Club, the latter was not prepared to organize offshore racing in European waters officially, and the task was taken over by a new club, the Ocean Racing Club.

By this time the Royal Cruising Club had some forty-five years of active life and experience behind it, and a set personality; not a few of its members disapproved of the idea of offshore racing, especially in the waters of Europe. But

At the start of the Bermuda race, 1923

the Cruising Club of America had no tradition yet; it had youth and iconoclasm. Yet when the idea of organizing ocean racing was mooted in the year after its establishment, the voice of opposition was strong, and for many years objections continued to be voiced. At first the voice of Frank Draper, a founder member, was the most forcible:

> 'Yacht racing does not need our aid.'
> 'I am of the opinion that the whole spirit of racing is radically opposed to the spirit of cruising. They are as oil and water.'
> 'Let us give our attention to cruising. Let the racing clubs attend to the racing.'
> 'From its nature, purpose and spirit the Cruising Club cannot be a racing club.'

Here is the authentic voice of the cruising man, the ideas that produced the Royal Cruising Club in 1880 advanced as fresh as ever in 1923. But, for better or worse, the Cruising Club of America turned later to sponsoring ocean racing, in the course of the next half century growing, for this reason, into a premier international yachting authority in what was to become – which it assuredly was not in 1923 – the most powerful branch of yachting. Did the Club belie its name and object in this activity? When the arguments were flying hotly during the early years there were those who, though they could not foresee the dominant position that offshore racing was to assume in international yachting during the later decades of the century, did insist that the encouragement of fast, efficient seagoing yachts was no less a function of seafaring under sail than navigation, meteorology and competent routine ship handling. And ocean racing was the way to produce such yachts.

Prior to 1923 there had been sailed five Bermuda races, the last of these in 1910 drawing only two entries. In 1923 the race was revived, and thus began the series which has continued until the present day with the event now drawing several hundred entries. The Cruising Club, maintaining its official attitude to racing, did not sponsor the 1923 race, but the organizing committee were all members and it appears that when the fleet of twenty-two yachts started off Sarah's Ledge on 12 June practically the whole membership were participants, including Frank Draper in his sloop *Flying Cloud.* Seventeen of the fleet were schooners, all but one of the yachts were gaff rigged, and as usual in the early offshore fleets some of the boats were badly fitted out and slow. All the boats finished, having sailed through strong south-westerly winds in the Gulf Stream. It was, inevitably, by later standards a slow race. The winner was an early yacht in the famous *Malabar* series, designed and owned by John Alden. This yacht architect from Boston was in the process of evolving a type of yacht that gave American offshore racing an advantage absent in Britain when the first ocean race in Europe was sailed two years later, and for some years afterwards. Derived from the New England fishing schooners and by stages refined in design, the Alden schooners were the fastest good seakeeping yachts of their size and period. There was nothing like them on the other side of the Atlantic.

In the following year it was voted at the annual meeting that the Club should organize but not sponsor the 1924 Bermuda, a resolution whose philosophical implications may not be quite clear. Entries for the race dropped to fourteen. In 1925 two events dimly prefigured the shape of things to come. A permanent club race committee was established, and John C. West, a young member of the Club went to Europe to take part in the first Fastnet race. He became one of the thirty-four founder members of the British Ocean Racing Club. In 1972 he was one of the few charter members still on the Club's list.

Weston Martyr, an Englishman living in New York, had sailed in the Bermuda races of 1923 and 1924. He returned home with an enthusiasm, though with few people to share it: 'I was shocked to discover that our yachtsmen had practically never heard of an ocean race and were apparently content to do their racing around short courses in sheltered water', he wrote later.

Weston Martyr had been many things since running away to sea in a square-rigged ship at the age of fifteen – a gold miner, a South Sea trader, a merchant and banker in the Far East, an infantryman in World War I, and he had adventured among head-hunters in Formosa and bootleggers in New York; he wrote of his experience in haunting prose. He now proceeded to bring ocean racing to Britain, writing himself on the subject and enlisting the help of George Martin and Malden Heckstall-Smith, which brought to the cause the influential platform of the *Yachting Monthly.*

As a result, in the early months of 1925 there was some talks in the clubs and notices in the national press about what was described as the 'Ocean Race'. The *Morning Post* (London) announced on 7 March that 'It has now been settled that the Ocean Race, of which preliminary notice has appeared in these columns, shall be sailed under the flag of the Royal Western Yacht Club, and that yachtsmen taking part in it shall be admitted to honorary membership for the time being'. The *Sunday Express* (London) followed with the information that 'A new 600-miles ocean race for yachts, starting from the Isle of Wight, has been arranged for the coming yachting season. It will be open to all nations belonging to the International Yacht Racing Union, and important entries are expected from America.'

So the first Fastnet race was announced. Its length was to be similar to that of the Bermuda race; the course radically different. The latter is a straight, purely offshore course from Newport to the low island and its coral reefs lying some 600

Sherman Hoyt, described on the dust-jacket of his memoirs as 'The Best Known Yachtsman in the World'. He was certainly one of the most experienced in both inshore and offshore racing.

miles to the south-east out in the deep mid-ocean. The Fastnet race course, as planned after much discussion in 1925 and unchanged in essential since then, lies for about half its length between the shores of the English Channel. Not only is the first long leg of the course, from the Needles to the Lizard, liable to be sailed hard on the wind, but the strong tides, the headlands with their overfalls, and the fickleness of Channel weather and visibility, provide such permutations of navigational possibilities that the wide ocean becomes an enviable simplicity in comparison. After the Lizard the seas lengthen, with the fetch of the Atlantic behind them, and the course lies approximately north-west on this ocean leg of the course out to the Fastnet Rock off Cape Clear, an area noted for severe conditions. Some of the criticisms were concerned with the dangers of sending small yachts into this locality. The return to the Longships may be easier, with the prevailing fair winds; but at the end of it there may be a landfall to make in low visibility on a lee shore. Such very briefly is the famous Fastnet course, now as rich in tales as the quest for the Golden Fleece, but in 1925 suggestive of no more than a surprising and possibly dangerous escapade proposed by nautical romantics.

Later Sherman Hoyt was to describe the Fastnet as '. . . the best and most sporting of all ocean race courses'. The Committee established to organize the race had the blessing of the Yacht Racing Association and was composed of experienced sailing people; but there were many criticisms, a few of them public.

Some from Claud Worth, who three years earlier had inspired Nutting with the idea of the Cruising Club, appeared in the London journal *The Field*; and the words of so competent and experienced a yacht seaman throw light on what was certainly the general attitude of the time towards a mode of yacht racing that was yet largely untried and could not be judged in the light of any existing experience. Claud Worth said:

'At the risk of making an unpopular suggestion, I venture to express a doubt which arises in my mind – are our latitudes suitable for a public ocean race? If two owners, experienced in ocean cruising, arrange a match involving several hundred miles of deep water, they know exactly what they are doing. But a public race might very well include some owners whose keenness is greater than their experience. If the weather should be bad, so long as there is a head wind they would probably come to no harm, for a good boat and sound gear will generally stand as much driving as the crew can put up with. But when running before anything approaching a gale of wind and a big sea in open water, conditions are very deceptive. A vessel of good shape and a reasonably long keel may run so easily and steadily that even an old hand, under cruising conditions, is apt to keep her running longer than is prudent. . . . I have more than once been compelled very reluctantly to heave-to and watch a fair wind running to waste, and have soon after had reason to be thankful that I was safely hove-to in good time.

'But if one had been racing one would probably have been tempted to carry on, knowing that some other competitor might take the risk. These conditions might not occur once in a dozen ocean races, but the magnitude of possible disaster should be taken into account.'

The measured words of Claud Worth, who taught so many men in his writings how to care for and handle their craft, found the indirect support of those with experience of racing only. Ocean racing was yet an activity on the fringe of yachting, beyond the experience of the last century during which modern yachting had evolved. That a gentleman does not go racing before breakfast or after tea was

Jolie Brise, a converted Le Havre pilot boat (see also p. 209) which belonged to Lt Commander E. G. Martin, first Commodore of the Royal Ocean Racing Club in 1925. The lines shown here were taken from the hull of the yacht. Length overall 56 ft, beam 15 ft 9 in, draught 10 ft 2 in, sail area 2,400 sq ft.

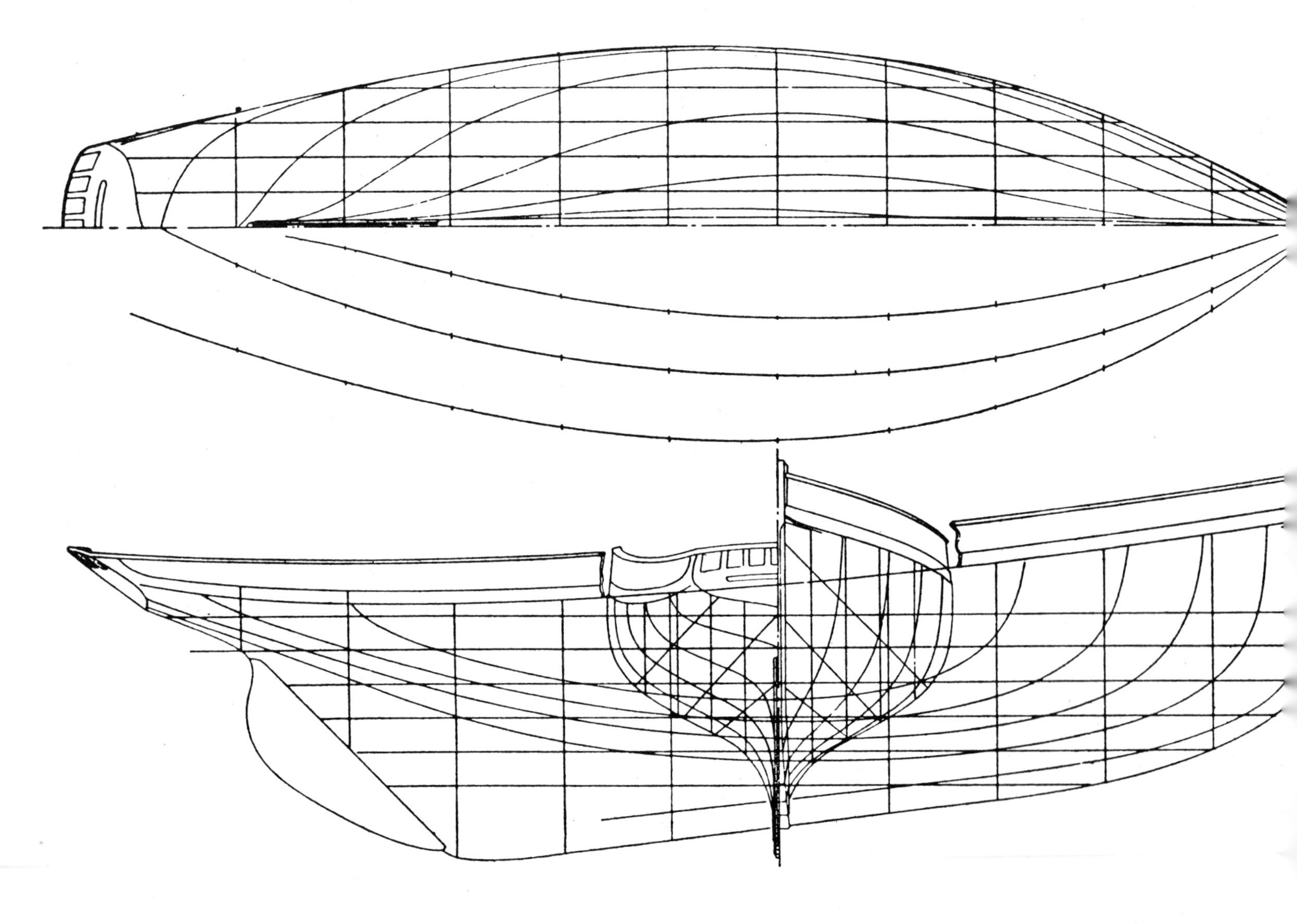

still being said many years after the first offshore racing had been organized in European waters. It was also argued that there could be little pleasure in racing during the dark hours which occupied much of an ocean race. The skill of the racing man had then to be subordinated to the prudence of the seaman, and the sport in the affair disappeared. There was also the fact, which was manifest, that the yachts taken out offshore racing in these early days were not what the dedicated racing man of the day could enjoy sailing. Those who have handled a thoroughbred acquire a taste not readily discarded. The ocean racing yachts of the period were hacks.

There was enough to be said against offshore racing at this time, when examined in the mirror of the past, to make its emergence as the premier form of yacht racing in craft bigger than the dayboat size, appear as no less than a revolution in taste, which was not complete until the years following World War II.

The entries for this first Fastnet were disappointing. The hoped-for American yacht did not materialize, and seven boats came to the starting-line. Their names and character should be recorded:

Jolie Brise: length overall 56 ft, length waterline 48 ft, beam 15 ft 9 in, draught 10 ft 2 in, sail area 2,400 sq ft. She was a Le Havre pilot boat built in 1913 by M. Paumelle, the celebrated designer and builder of this class of vessel.

Jessie L: length overall 42 ft, length waterline 37 ft, beam 13 ft 4 in, draught 6 ft 6 in, sail area 1,250 sq ft. A Bristol Channel pilot cutter built by Hambly as early as 1887, but now Bermudian rigged.

Saladin: length overall 49 ft, length waterline 43 ft, beam 14 ft 8 in, draught 8 ft. She also was a converted Bristol Channel pilot cutter.

North Star: length overall 49 ft, length waterline 46 ft, beam 14 ft 6 in, draught 7 ft 6 in. This yacht was a Norwegian Colin Archer type of ketch built in 1925 at Tvedestrand.

Banba IV: length overall 38 ft, length waterline 34 ft, beam 12 ft, draught 7 ft, sail area 1,330 sq ft. She was also of the Norwegian type, but designed by Frederick Shepherd, and of deeper section than the true Norwegian model.

Fulmar: length overall 38 ft, length waterline 34 ft, beam 12 ft, draught 7 ft, sail area 1,050 sq ft. She was one of the two boats in the fleet built as a yacht, but she was on working-boat lines, built in the West Country in 1901.

Gull: length overall 44 ft, length waterline 37 ft, beam 10 ft, draught 7 ft, sail area 1,597 sq ft. A typical, narrow Camper and Nicholson cutter, stemming from the plank on edge tradition of design, though with its more extreme features modified. She was built in 1896, and was therefore the second oldest yacht in the fleet.

The fleet thus consisted of three converted pilot cutters, two Norwegian type yachts derived from working boats, a vessel built as a yacht on working boat lines nearly a quarter of a century before the race, and a cruiser built as a yacht by the leading British firm of yacht designers and builders *more* than a quarter of a century before the race. *Fulmar*, belonging to the Royal Engineers Yacht Club, and built in 1891, was still racing in 1955, when she won Class I of the Clyde Cruising Club's closing regatta: 'And now in age I bud again'.

It transpired to be a relatively easy race. E. G. Martin in *Jolie Brise* – shortly to be the first Commodore of the Ocean Racing Club – carried light winds almost the whole way to the Fastnet, which he rounded first, followed only forty minutes and seventy-five minutes later respectively by *Gull* and *Saladin*. The Rock was rounded with a falling glass and westerly wind that backed to the south-west and increased, sending the leaders spinning home on a reach. *Jolie Brise*, first across the line and winner on corrected time, completed the course in some six days and three hours. Three boats retired. One of them, *North Star*, did so within a few cables of the finishing-line off Drake Island, where light wind and a foul tide held her back, and having no hope of winning she started her engine. *Fulmar* was second on corrected time, *Gull* third.

Following the race the organizing committee transformed itself into the Ocean Racing Club. Thus after one such race only Britain possessed an organizing body for the young sport which the U.S.A., despite its head start in the activity, still lacked owing to the equivocal attitude of the Cruising Club to racing. In later years the Royal Ocean Racing Club – it received the title 'Royal' in 1930 – with its London West End clubhouse and large American membership became the only permanent headquarters of ocean racing, the Cruising Club never having had a building of its own.

The rules of the new club are revealing. It was to organize a race annually of not less than 600 miles round the Fastnet Rock, and membership was to be confined to those who had taken part in the race. Herbert Stone, another early American member, was however elected to the club without this qualification. Time allowance for the race was not to be allocated crudely on form but upon the basis of a rating formula; and thus emerged upon the stage of yachting

Tally Ho, winner of the third Fastnet race in 1927 in very heavy weather, when thirteen out of fifteen yachts were forced to retire.

La Goleta, one of the two American entries in the 1927 Fastnet race, and the only yacht apart from *Tally Ho* to finish the course.

history what later became so well known as the 'R.O.R.C. Rule'. In this respect too Britain led the U.S.A., for the Cruising Club did not place measurement and rating on a scientific foundation until 1934. But if Britain led the way in administrational and technical organization, the later 1920s and 1930s were to show the superiority of American yachts and racing techniques.

The next year produced the first of what may be called 'classic' Fastnets, the outcome of bad weather and two American entries. These, *Nicanor* and *La*

Goleta, were both of the Alden semi-fisherman type schooners. Of the fifteen starters two alone completed the course, which ended in a stirring duel between the British cutter *Tallo Ho* and *La Goleta*. The brilliant series of American successes in Fastnet racing of 1931, 1933 and 1935 (the race becoming biennial after 1931) was anticipated in *La Goleta*'s 1927 performance.

A dismal start in heavy rain led the fleet on to beat dead to windward down Channel in rising wind. Three days later the fleet was reduced to six yachts, sheltering under the Lizard. *Jolie Brise*, out ahead, found conditions too bad, and off the Manacles she retired also, to be joined later by *Nicanor*, which gybed and broke her gaff on the way out to the Rock. The heavy weather race eventually reduced itself to the 47-ft cutter *Tally Ho* and the 54-ft schooner *La Goleta*, the former a well tested boat, the latter fresh from the builder's hands and leaking abominably. Lord Stalbridge, owner and skipper of *Tally Ho*, wrote of conditions off the Lizard: 'I saw an extraordinary sight; a big oil tanker was steaming into it and as she lifted we could see her keel from forefoot to well abaft her foremast and then, as she dipped, her propeller and practically the whole of her rudder came clear out of the water. This will give you some idea of the sea that was running.'

Then the two boats were crossing the Irish Sea in a wind that had moderated and freed, making it possible to lay the course to the Fastnet. Under these conditions *La Goleta*, initially well astern, gained the advantage of her length, coming up over the long swell and passing ahead under the lee of the cutter. But it was *Tally Ho*, in almost a calm, which drifted first round the Rock. The yachts were in the centre of the depression.

When the wind came again England was a lee shore, the seas at times very heavy with wind against tide. Rounding the tip of Cornwall later, after periods hove-to, *La Goleta* was the first to gain the more moderate seas under the land. She crossed the line off Drake's Island first, but was not enough in the lead to save her time on *Tally Ho*, which became the winner of the third Fastnet. The number of boats that had retired caused some unfavourable comment, but even amongst those with good knowledge, a note of taut drama was heard concerning the dangers of the race: 'I almost wept for joy when I heard that the heroic boys, who manned the Ocean Racing Fleet, had reported with numbers intact. . . . And well I know that you also must have suffered mental anguish while their fate was in doubt and their respective harbours of refuge unknown.' This was written to the Secretary, Malden Heckstall-Smith, by William McC. Meek, Surveyor of Yachts for *Lloyd's Register* and a valued official of the Club. On the other hand were the objections to the mass retirements. Both attitudes are a measure of the weakness in offshore racing at the time – the inexperience of the crews, the inadequacies of the boats.

Two yachts stand out in the early story of European offshore racing. The Le Havre pilot boat *Jolie Brise*, winner of the Fastnet in 1925, 1929 and 1930, and owned by the first Commodore of the Club, holds a unique place in the Club's history, indicated by the model of her standing in the Fastnet room of the clubhouse in London. On another table is a model of a remarkably different type of yacht, the American schooner *Nina*, which was new when *Jolie Brise* was in her fourteenth year and stood for the future while the older boat was an expression of garnered wisdom unwillingly becoming out of date. Paul Hammond's *Nina* came to Europe for the 1928 Fastnet – came, saw and conquered. It was a race which brought a high proportion of windward work and calms. *Nina*, stepping out almost like an inshore racing class yacht held an all too clear mirror to the indifferent sailing qualities of the typical ocean racers of the time, and she

Stages in design for ocean racing, as shown by models in the Royal Ocean Racing Club, London. *Jolie Brise* (right) stands beside the American schooner *Nina*, which won the Fastnet race in 1928 and represented an early stage in the development of the specialized ocean-racing yacht.

crossed the finishing-line nine and a half hours ahead of the other American entry *Mohawk*. But *Jolie Brise* saved her time on *Mohawk* and was placed second. Included in the racing fleet this year, incidentally, was an inshore racing yacht, the Anker-designed 12-Metre *Noreen*, with a freak Bermudian rig having the mast stepped abaft amidships and a great show of headsails. She retired. And for the first time there was a French entry, Leon Diot's *L'Oiseau Bleu*.

But it was *Nina* that focused attention – *Nina* which had wiped up the whole fleet with a graceful gesture, though lacking, owing to misadventure, either chronometer time or a reliable log. She was in the tradition of the American schooner, but the tradition had been stretched. She suggested, but pretty vaguely, the semi-fisherman type. It had been refined and polished to produce a dedicated ocean-racing machine, designed to fit the rating rule, and with an extreme type of schooner rig, the mainmast nearly amidships, the foremast short and well into the bow. She was a boat of great expense and complication, unsuitable for cruising in European waters, and she needed a large crew for her length of 59 ft and sail area of 2,500 sq ft.

The outcome of this Fastnet was not happy. There was much adverse comment on *Nina*, not a little of it well publicized, and one member of the committee considered that this type of boat should be barred from the Fastnet; which would have been tantamount to strangling at birth the real ocean racer. As the present author wrote in the history *British Ocean Racing*: 'National acerbities were aroused between Britain and the U.S.A., and we cannot do other than confess that it was entirely the fault of the former.' Sherman Hoyt wrote a spirited defence of *Nina*, and pulled no punches over what he regarded as the unsporting attitude of the British. The Committee of the Ocean Racing Club made a riposte that was almost a stroke of genius, the secretary replying:

'I have been asked by my Flag Officers to inform you that you have been unanimously elected Rear Commodore of the Ocean Racing Club at a recent General Meeting held on 17 December.'

The maligned *Nina* had been designed to the rule governing the Spanish Transatlantic race of 1928. In the smaller of the two classes for the race *Nina* was the largest and the only highly developed ocean racer of the emerging fashion. Against her were two Alden-designed semi-fisherman type schooners, *Pinta* and *Mohawk*, both entries in the previous year's Fastnet, the former then named *Nicanor*. The three raced across the Atlantic on a handicap basis for the first time. The Big Class, headed by the magnificent three-masted schooner *Atlantic*, rich still in the memory of her passage across the Atlantic in twelve days, four hours and one minute between Sandy Hook and the Needles in the Spanish race of 1905. Now nearly a quarter of a century later, she brought to the new kind of ocean racing, in small craft crewed by amateurs, the echo of the spacious past when large yachts were handled by focsles full of paid hands. Her closest rivals in the Big Class, which started a week after the smaller, was the Herreshoff-designed racing schooner *Elina*. She, like *Nina*, was favoured by the hard windward work towards the end of the race, which enabled her to defeat *Atlantic*, and *Nina* to prove herself peerless against the gaff-rigged semi-fisherman type. The race was the herald for the eclipse of the latter in the first-class ocean racing. The Spanish welcome for the American fleet at Santander was in the grand manner of the *Atlantic* herself. It was from these splendid festivities that the crews of *Nina* and *Mohawk* tore themselves in order to sail north to the Fastnet race – an

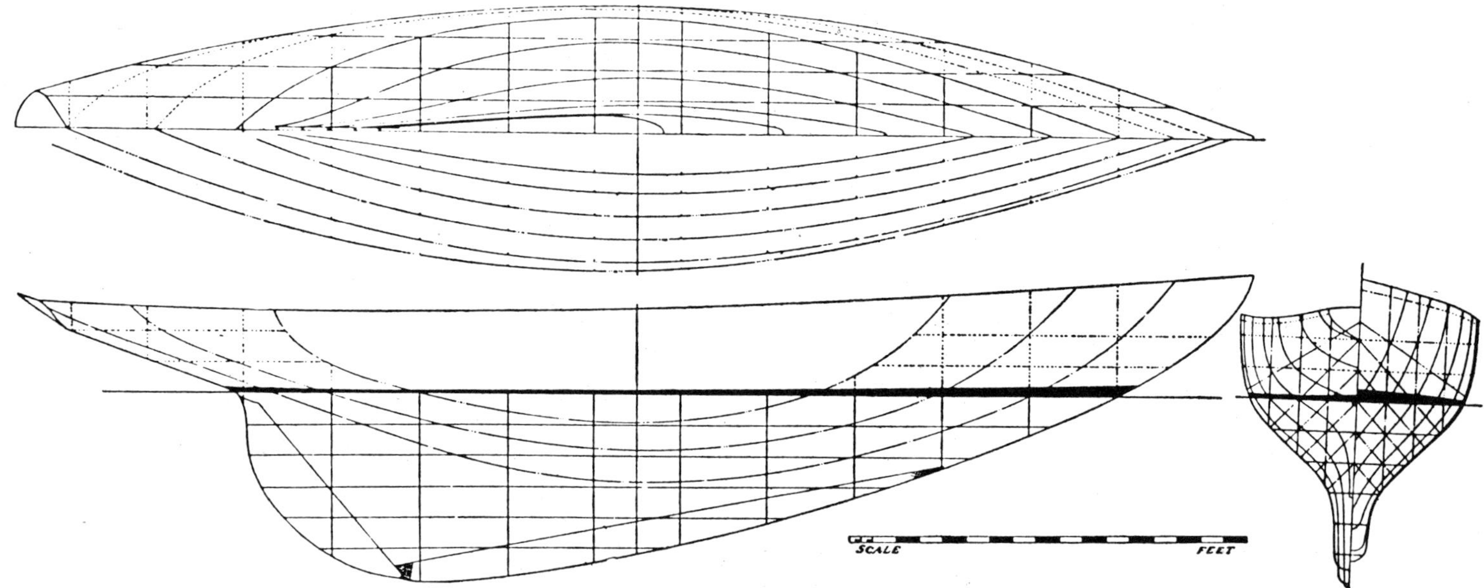

The lines of *Dorade*, designed by Olin Stephens, showing the narrow, easy form of the hull.

additional reason for the resentment felt at the criticism loaded on *Nina* at the end of the race. With the Bermuda race drawing twenty-four entries in that year, then the Transatlantic – the first since 1905 – followed by entries in the Fastnet, it was the most active year yet experienced in American ocean racing.

The Bermuda race, like the Fastnet, had become established as biennial, the two annually alternating, an arrangement that was to become the fixed pattern of international offshore racing until the present day. Also casting long shadows towards the present, though little noticed at the time, was the victory of a little 30 ft sloop named *Kalmia* in a race organized by the Gibson Island Yacht Club in Chesapeake Bay from New London to Gibson Island. The record of the race might have become mislaid down the winding ways of history but for the fact that the little sloop had been designed by Olin Stephens.

The Golconda of Wall Street lay in ruins when the next Bermuda was sailed

Alfred Loomis in his younger days.

in 1930, despite which there was an entry of forty-two boats, far greater than in any ocean race up to this time, and a precursor of the future. Another was the appearance in this race of the 52 ft overall Bermudian yawl *Dorade*. The Alden-designed *Malabar X* was the winner in Class A. *Dorade*, destined to become one of the most influential yachts in the story of ocean racing, finished modestly second in Class B to the small veteran yawl *Malay*, which was able to save her time in a bunched-up finish of the race.

In 1931 a transatlantic race from Newport to Plymouth was the prelude to the Fastnet and the first occasion when Ocean Racing Club and Cruising Club jointly organized a race. The ten yachts which set off from Brenton Reef on 4 July in a warm light wind amid a circling crown of sightseeing craft represented the most impressive display yet of confident ocean-going in relatively small amateur-manned yachts. The Atlantic was being taken in their stride. This race was won by *Dorade* and brought her to the line for the Fastnet race, which she also won. The next transatlantic race was in 1935, and the winner, *Stormy Weather*, which like *Dorade* was an Olin Stephens-designed yacht, also won the Fastnet of that year, the third Fastnet victory for a yacht by this architect.

Between 1931, when the largest Fastnet up to this time was sailed, and 1935, when an equally strong fleet appeared, British ocean racing almost became defunct. Despite the convincing example of such yachts as *Nina*, *Dorade*, *Landfall* and *Highland Light*, no pure bred ocean-racing yacht had yet appeared in Britain. In the Cruising Club there was continuously the dichotomy in attitude of the cruising and racing members. This could not occur in the Royal Ocean Racing Club. Here it was rather that the members, all supporters of offshore racing, were yet wedded to the accepted form of the British cruising yacht and the seagoing modes of the cruiser rather than the newly evolving techniques of the racing man as adapted for offshore work.

The yachting press of the period examined with tart comments on the apathetic situation, the smallness of the fleet that regularly raced offshore, the modest roll call of the regulars, and some sections of it were frankly critical of the successful American yachts, which it was claimed were not adequate cruising yachts – and had not the Royal Ocean Racing Club been founded to provide *bona fide cruisers*, not man-eating racing machines sacrificing comfort and even safety for speed? A great deal of nonsense was printed, and Alf Loomis said a little wearily: 'One of these years England is going to win the Fastnet with a yacht that satisfies every requirement of comfort lovers and yet expresses the modern need to keep moving through the water when close-hauled to wind.'

She was going to achieve no less than this shortly, though a little laggardly in doing so. Meanwhile it appeared that there might be only one British entry for the 1933 Fastnet. Weston Martyr hurled the most readable abuse in the press at this backwardness, and eventually three boats came to the line to meet three Americans. It was a light wind race, with the honours pre-eminently American. The exception was provided by *Flame*, now belonging to the well-known yacht builder and architect Charles Nicholson, who had designed and built the last three America's Cup challengers. *Flame* was not derived from the working-boat type but was a narrow, heavy displacement cutter of 1900, typical of English yachting developments in the most advanced cruising developments of her date. She was now Bermudian rigged, but also rigged for a heavy weather race. She crossed the finishing line first, but the smaller *Dorade* had an easy win on time allowance. *Dorade* won the race for the second time, *Flame* dropped to third on time allowance.

SUMMARY OF R.O.R.C. RACES 1925–39

1925	1926	1927	1928	1929	1930	1931	1932	1933
Fastnet	Fastnet	Fastnet	Fastnet Channel	Fastnet Channel Santander	Fastnet Channel Santander Dinard	Fastnet Channel Transatlantic Dinard Haaks Maas	Channel Dinard Haaks Maas	Fastnet Channel Dinard Heligoland Maas

TOTAL RACES IN SEASON

1925	1926	1927	1928	1929	1930	1931	1932	1933
1	1	1	2	3	4	6	4	5

TOTAL RACES SAILED 1925–39 = 70

While ocean racing in Europe thus appeared to be a fire guttering soon after it had been lit, the Bermuda races of 1932 and 1934, which bracketed the unfortunate Fastnet of 1933, presented a picture of virility which defied the prevailing depression. There were twenty-seven entries for the race in 1932 and twenty-nine in 1934. Among the fleets were such boats of the new era of ocean racing as the schooner *Grenadier*, in the tradition of the Alden fisherman type, *Vamarie*, a wishbone ketch which was first across the finishing-line in 1934; also the sloop *Edlu*, which won on corrected time in that year, and *Dorade* and *Stormy Weather*, the boats directly pointing the way to the future. They were distinguished fleets rich in the spirit of the new ocean racing era, with which Europe was not yet in step. But while the technique was weak in Europe, two English yachts were entries for the 1932 Bermuda race; *Jolie Brise* was there, the European pioneer, now owned by Robert Somerset, who later owned briefly the American *Nina*, and *Lexia*, a recently built gaff-rigged cruising cutter in the typically British conservative manner. One of the boats belonged to a past well loved, the other to a still respected tradition of seamanship.

The table above illustrates the growth of European offshore racing during the inter-war years. It opens with the first Fastnet race; moves through the years when the Royal Ocean Racing Club was organizing increasing numbers of races annually – six in 1931 – on to 1933 when, despite four other races being held, the Fastnet almost failed; then from 1934 onwards a period of more confident growth.

The high-water mark of inter-war offshore racing in Europe was reached in the 1937 Fastnet race. Twenty-eight yachts entered from five nations: eighteen British yachts, seven German, a Frenchman, a Dutchman and an American. How far away already 1933 seemed with its three British entries. It was the finest

1934	1935	1936	1937	1938	1939
Channel Dinard Maas Burnham-on-Crouch–Heligoland Heligoland–Copenhagen Plymouth–Belle Île	Fastnet Channel Dinard Heligoland Plymouth–Belle Île	Channel Dinard Maas Heligoland Plymouth–Benodet Falmouth–Clyde	Fastnet Channel Dinard Heligoland Maas Southsea–Brixham Ijmuiden–Solent La Baule	Channel Dinard Heligoland Maas Dover–Kristiansand Kristiansand–Copenhagen Copenhagen–Warnemunde Falmouth–Kingstown Kingstown–Clyde Brehat–Brixham	Fastnet Channel Dinard Burnham–Weser La Rochelle Southsea–Brixham Harwich–Solent Solent–Falmouth
TOTAL RACES IN SEASON					
6	5	6	8	10	8

A further stage in design for ocean racing, represented by the American yawl *Dorade*, a relatively small yacht, with an overall length of 52 ft and waterline length of 37 ft 3 in, by the standards of the early 1930s. She won the Fastnet races of 1930 and 1931 with impressive ease.

ocean racing fleet in quality and size yet to have been started outside American waters, and more internationally representative than any. *Saladin* and *Banba* were of it, veterans now of the first Fastnet, amongst the new kind of offshore racing yachts which were the favourites for the race, but had not been a line on paper when the two old stagers had first set their pioneering course for the Rock. In a race which provided little weight of wind, and some of which was a light wind struggle, the Dutch *Zeearend* was winner on corrected time after crossing the finishing-line within the same minute as the British *Bloodhound*, which dropped to fourth place.

That Fastnet fleet, impressive by the standards of 1937, was by no means enough to vouchsafe, by its size and international composition, any clear vision of the future, when not twenty-eight but almost 300 yachts, from not five but nineteen nations would cross the Fastnet starting-line. And could that vision have been granted it might have alarmed rather than inspired. When the vision did materialize there were those who could feel misgivings at the pressure of the later twentieth-century citizens' rush to the ocean. What had been an idiosyncratic sport had reached the proportions of a social movement reflecting ambiguously on the conditions of life on shore.

In the narrower world of yachting the modes of offshore racing changed, of which there were plentiful signs prior to 1939. The Royal Ocean Racing Club, armed with its measurement rule and system of time allowance, was tending to become the guardian of all forms of handicap racing, including the short port-to-port passage races that were in no sense offshore events.

Below

Fastnet race 1938. The yacht in the foreground is *Golden Dragon*.

Opposite

Britannia painted by Norman Wilkinson in the early 1930s, late in the yacht's career.

K 1

Shamrock V. Painting by Tom Ludlam, 1930.

The last Fastnet before the outbreak of war was the eleventh; and the race had become a yachting institution. There were German and Dutch entries. In the same year there was embarrassment about the Weser race, which had been substituted for the Heligoland race. The present author wrote this about it in *British Ocean Racing*:

'The problems encountered in organizing the Weser race seem today like the first dropping of the curtain on the pre-war phase of British ocean racing. At the end of April it was discussed whether the race ought not to be cancelled. Feeling against Germany was so high that there was danger of incidents if German yachtsmen appeared in Harwich; a fact that astonished them when Ray Barrett, in what must have been a most embarrassing interview with Dr Perlia and Herr R. Schmidt, told them so.

'The matter was considered at a Flag Officers' meeting on 27 April. Opinions were divided. Some felt that it was better to risk a few incidents, which anyhow would not be caused by yachtsmen, than for the Club to initiate the rebuff of cancelling the race. The point was made that if the race were to be run it was important for the decision to be conveyed to the Germans before a speech that was due from Hitler on the following day. Then, if necessary, the speech might be made the excuse for cancellation. A cable was immediately sent to this effect.

'Hitler, on this occasion, did not manage to disturb the race, or any others, and the season ran its full if uneasy course. There were Fastnet, Channel, La Rochelle races; the Weser race, and the shorter events – St Malo–Dinard, Southsea–Brixham, Harwich–Solent.'

War came to Europe with ocean racing well established on both sides of the Atlantic. For much of the period between the wars, while the sport advanced from being a fringe activity to one of central importance, Europe had been the learner, the U.S.A. the teacher. For a few years, as we have seen, there seemed the possibility that ocean racing in Europe might die from lack of even the slender support it had initially gained.

Now not only had the Royal Ocean Racing Club survived; it was powerful. Curiously, it was in 1939 that the older Cruising Club of America burst again into schism, from which the Royal Ocean Racing Club, with its single object of fostering ocean racing, was free. Should the Cruising Club continue to sponsor ocean racing? There were suggestions that an Ocean *Racing* Club of America should be founded, leaving the Cruising Club to fulfil its original purpose. A committee was organized to thrash out this enduring question. The majority report of the Committee recognized that professionalism and an element of over-enthusiasm for distinguished racing reputations existed among the Club membership; as a minority report expressed it, 'The cruising men neither understand nor enjoy the racing men's philosophy . . . cruising members are becoming fearful of eventual domination of the Club by racing men.'

It was this undoubted professionalism in ocean racing that had given the U.S.A. its lead. In Europe, despite a club devoted exclusively to ocean racing, and which was an undivided one, it was lack of professionalism that retarded the purer technical skills of the sport. Which was the better mode? There have always been a few who question the value of too much enthusiasm, in the sense that Tallyrand meant it when he prided himself that his assistants in his foreign ministry quite lacked the quality. The Cruising Club of America in the majority report favoured continuing to sponsor the Bermuda race.

Part IV

Twentieth century explosion from 1946

13 Re-establishment again

By August six years later the coastal waters were putting on their sails again. Small area after area, nicely defined on the charts, was opened to pleasure craft after five summers of prohibition. Every private boat out sailing was a small triumph for peace. In Britain there was much conjecture about the disposal by the government of surplus craft built during the war but suitable for conversion into pleasure craft; such might be the one means of owning a boat at moderate cost in a world so desperately short of materials. Yet the yacht yards, completing the hang-over of war orders, were by the spring of 1946 reverting to yachts. Early in the same year it was expected that there might be twenty entries for the revived Bermuda race; and the great names of the 1930s emerged from retirement – Henry C. Taylor's *Baruna*, George E. Roosevelt's *Mistress*, Briggs Cunningham's *Brilliant*, Rod Stephens' *Revenoc*. It seemed far away old times that the names recalled to the brave new world of peace. An English yachting editor noted that the applications for craft being disposed of by the government much exceeded the supply; he considered this a cheerful augury for the future of yacht and boat building, while someone wildly prophesied that *marinas* might become necessary in Britain. The word struck a discord in English ears at this time; people suggested 'yacht harbour' as an alternative.

In November 1946 the first post-war meeting of the International Yacht Racing Union was held in the new offices of the Yacht Racing Association in London, and the following countries were reported as being still actively associated with the union: Denmark, France, Britain, Switzerland, Spain, Czechoslovakia, Holland, Belgium, Finland, Brazil, Hungary, Sweden, Portugal, Norway, Argentine, Cuba, Uruguay. Those responsible for the reorganization of international yacht racing were imbued with one far-seeing idea which was expressed when Crown Prince Olaf of Norway read a report to the conference: 'The Scandinavian Yachting Union, therefore, believes that the time has come to revise the rating rule so as to produce yachts of a more seaworthy and cruising type; in other words to have cruisers that can race rather than racers that can be used for cruising '

The rating rule objected to was the International Yacht Racing Union rule of the inter-war period which had produced the 6-, 8- and 12-Metre classes. The view was generally held that in the impoverished world left in the trail of the fighting these lovely specialized machines, which in the 8-Metre and 12-Metre classes were big enough to contain comfortable accommodation, but never did so being dedicated to speed, should be replaced by smaller dual purpose yachts, adequately rigged and equipped for seagoing. Such craft, unlike the pre-war Eights and Twelves, would retain their value when their first-class racing days were over. The Permanent Committee of the International Yacht Racing Union therefore stated in its minutes 'Recommended that the International Rule of Measurement be altered and that three years' notice be given to that effect. It is proposed to appoint an International Sub Committee of Technical Advisers to draft a new International rule with reference to the above mentioned proposal, and to invite the North American Yacht Racing Union to be represented on that sub-committee.'

For three years the sub-committee proceeded with this task while the day of the cruiser-racer spread over the seas of yachting in the fleets of the Royal Ocean Racing Club and the Cruising Club of America. The rules of both these bodies were examined as possibilities for adoption by the International Yacht Racing Union, but eventually it was a formulation basically due to the Norwegian architect Barne Aas that was adopted. It provided for five classes, to be known as the 'International Cruiser-Racer Classes', the ratings being 7-, 8-, 9-, 10½- and 12-Metres. The yachts were smaller and tougher for their rating than those of the pre-war International classes; the 8-Metre Cruiser-Racer was shorter and, proportionately beamier than the pre-war 8-Metre, had a smaller, sturdier rig and rules enforcing a very much higher standard of accommodation. More of the total weight was in this and in the hull structure, and less in the keel; the result was a less perfect regatta racing machine but a more versatile yacht.

This new International rule, which came out in 1950 to replace the former one that had produced the great pre-war regatta fleets, failed to gain a comparable distinction either for regatta racing or offshore. A class of 8-Metre Cruiser-Racers grew up on the Clyde. In Scandinavia, the birthplace of the rule, it had a limited use, and also in France. By the time the rule came into force those of the Royal Ocean Racing Club and Cruising Club of America had undergone swift and pregnant developments during the first enthusiastic seasons of post-war yachting; perhaps these now were doing better what it had been intended the International Cruiser-Racer rules should do. It was also a handicap to the success of the latter that not only did the boats produced tend not to fare too well when racing with the offshore fleets under their respective rules, but the latter had also begun to dominate the regatta racing with their time allowance system.

A new International rule was also produced to replace the 6-Metres. Hitherto, as we have seen, the three International Metre classes of 6-, 8- and 12-Metres, were measured by the same rule, though the two latter were yachts with accommodation and the first without. This involved a questionable principle. The 6-Metres were now coming in for harsh criticism which tended to discount and forget the many years of magnificent international racing, the profound developments in race sailing technique, that the class had engendered. Now they were abnormally expensive and in the harsh aftermath of war thoroughbred qualities could appear as slightly reprehensible. Charles Nicholson himself, who had much responsibility for the rule governing the class, went so far as describing it as one of the worst misjudgements of his professional life. The 6-Metres were relatively heavy craft for what they were; which in effect were open boats. With hulls of light construction, though to Lloyd's racing rules and beautiful examples of the shipwrights' art, and with no accommodation, their narrow hulls were expensively lead-loaded, some seventy per cent of the total weight being in the keel. Above the hull was a large rig by the standards of the day spread on light mast and spars. The very extravagance of their features made them the superb sailing machines they had become, and now assured their extinction.

In their place, and to a quite differently formulated rule, came the International Yacht Racing Union's 5½-Metres. Shorter, proportionately very much lighter, with less sail area, originally measured in such a way as to preclude the

	$5\frac{1}{2}$-Metre	6-Metre
Length overall	35 ft 6 in	38 ft 6 in
Length waterline	22 ft 0 in	23 ft 6 in
Beam	6 ft 4 in	6 ft 1 in
Draught	4 ft 4 in	5 ft 6 in
Displacement	1.9 tons	4.1 tons
Sail area (measured)	300 sq ft	452 sq ft

genoa – a piece of economically inspired legislation which became most unpopular in many countries in which the class was raced. A comparison between a $5\frac{1}{2}$-Metre and a 6-Metre appears in table, left.

The rule was not wholly new when it was adopted in 1950, being an adaption of a basic formulation that had already done great service. It had been produced for the Boat Racing Association in 1912, a body long defunct. It had briefly governed an International 18-ft class in 1920, a class which, significantly in the light of these later events, had been driven out of existence by the 6-Metres which was now killed by a class to 18-ft rule. It has been in continuous use, in a

Above

The American 6-Metre *Goose* (left) racing against the British *Johan* in the Solent.

Left

The post World War II International 5.5-Metre class.

slightly modified form, in the model yacht racing 'A' class between 1922 and the present day. And the basis of the Royal Ocean Racing Club's famous measurement rule, which was to dominate European yacht racing in the years following 1945, was an adaption of the same rule.

In due course the $5\frac{1}{2}$-Metres became one of the Olympic Games yachting classes. It gained a good following in Scandinavia and was sailed in Bermuda and some European countries. There has never been more than a small number in Britain. The merit of the rule and the boats have failed to raise it to the position of a worthy successor to the 6-Metre class.

Despite being criticized, the 6-Metre class was not readily dismissed by the replacement $5\frac{1}{2}$-Metres. Smallest of the pre-war International Yacht Racing Union's inshore racing craft, it was now the largest – nobody had yet contemplated the resurrection of the 12-Metres for America's Cup racing – though where they had once raced in scores they did so now in little exclusive handfuls. They were described by one admirer of them as 'a yacht that has, despite its faults, become the symbol of the best that any nation can produce'.

In 1949 the Americans won the third British–American Cup outright. The Seawanhaka Corinthian Yacht Club then presented a fourth cup in memory of their ex-Commodore, the late George Nichols. In 1951 a team of the same three boats, *Llanoria*, *Goose* and *Firecracker*, that had taken away the third cup two years earlier came to the Solent to defend the fourth, against the British Sixes *Circe, Johan* and *Marletta*, the two former being pre-war boats and *Circe* and *Marletta* being designed by David Boyd, who some years hence was to design two America's Cup challengers.

The latter were to be notably unsuccessful, but in the smaller class racing competition was interesting. By the end of the fourth race in 1951 the British team of Sixes had won three; another win would gain the cup. The poor performance of the usually fast *Goose* had contributed to this situation; and it was learned that

Seawanhaka cup racing, a windward leg, in 1953.

her helmsman was suffering from an illness shortly to prove fatal. In other hands *Goose* returned to form. The Americans won the next two races, gaining the first three places in the second of these, and then the first two in the following one, which brought the team four wins and the cup. America won again two years later in the U.S.A., and took also the Seawanhaka and One Ton Cups and a new event for the 6-Metre class known as the New World Against The Old. American superiority in design had become strikingly evident, and much of it might have been contributed to the tank testing now so highly developed in the model tank at Hoboken. Indeed, Dr Kenneth Davidson who had evolved the technique for testing model sailing yachts, and was himself a keen yachtsman, had confessed that his interest in tank testing was then centred on the 6-Metres. This may be regarded as a warm tribute to the merits of the class.

Four new Sixes were built in Britain for the 1955 season, when an American team came to the Solent to defend the British–American Cup, which they did with readiness in four straight races, *Llanoria* winning every race in the series. But now, after serving generations of racing people since the beginning of the century, at last the 6-Metres were approaching their sunset. The day was nearly done for one of the greatest of all yacht racing classes in the history of the sport.

Yachting, sailing, motor yachting, just boating – the terms were becoming confused and employed uncertainly in the common speech – was extending its place in the social lives of the nations. In Britain for some years there had been a movement favouring the creation of a national representative organization for yachting. There was a growing need for a responsible body empowered to deal with the boat building industry, dock and harbour boards, local government authorities, even with Whitehall. To the older fashion outlook the idea of such a body was faintly abhorrent, something inimical to the salty freedom, the fine privacy of those who use the sea; an attitude succinctly expressed by one correspondent to a yachting journal: 'Re national representative association for yachtsmen – my God!' The editor, representing the contrary view, published the letter under the heading 'Unconstructive invocation of the Deity'.

A solution, and the one ultimately adopted, was to broaden the constitution of the Yacht Racing Association, whose function was already rather wider than its title might indicate. A step in this direction was taken when the Council of the Association approved the formation of a General Purposes Committee, on which sat representatives of yacht and motor yacht cruising interests and dealt with such matters as mooring charges and the growing number of legal and administrative matters affecting yachtsmen. Following this step, taken in 1947, the Yacht Racing Association gave expression to its expanding role by changing its title to Yachting Association, and a year later, in 1953, it received the title of the Royal Yachting Association. It became what the Yacht Racing Association had sometimes in the past been called, contrary to the opinion of a high proportion of cruising people – a body of distinction and utility. The Association experienced a growth with that of yachting. It had 312 member clubs and 235 individual members a few years before the formation of the General Purposes Committee; in the year it became the Royal Yachting Association it had 650 member clubs and 2,100 private members; by the mid 1960s the two categories of membership had increased to 1,356 clubs and 22,000 individual members, the latter being not much short of a one hundred fold growth since 1945.

At the present author's request, in 1955 Robert N. Bavier Jr wrote a brief review of the American world of yachting. Some ten post-war years had by then passed. The spirit of recreational activities afloat, which had emerged in an almost frenzied reaction against the war years to spread over the coastal and inland waters of so much of the world, had become a firmly established mode of life among many different peoples. Nowhere was the quintessence of matter revealed more obviously than in the U.S.A. Yachting had become transformed; democracy had raised new permutations in an activity old and often exclusive. Some twenty more years have passed since 1955, to spread further and amplify what we knew then on the waters of the world where people take their pleasure.

'The Americans', said Bavier, 'have taken to yachting as never before. More than 200 million Americans went afloat (in 1955) in five and a half million boats of some sort or other, from outboard-powered rowing boats to ocean racing auxiliaries. The boating boom was evident in such tradition-soaked centres as Long Island Sound, Marblehead, Newport Beach, California and Seattle, Washington; but these centres now share the yachting stage with almost all sections of the country wherever there is a lake, bay or broad river. All America is getting into the act.'

Bavier proceeded to emphasize the unequalled facilities the U.S.A. is able to offer those who wish to take their pleasure on the water; a fact able to raise envy amongst those elsewhere who were experiencing a pressure of boat population in this new phase of yachting – a pressure ever increasing since: 'It may be hard to comprehend the geographical extent of yachting in America. The wealth and variety of the cruising grounds has certainly contributed to the growth in boating. In summer months for eastern yachtsmen, cruising in New England offers hundreds of miles of scenic waters with literally thousands of harbours to choose from. The sport is active all the year round farther south and as winter comes many migrate down the coast, following the season to Florida and Gulf waters. The intra-coastal waterway permits boats with power to cruise from Massachusetts to Florida with only a score of miles of outside seaway for the entire route.

'The Great Lakes alone offer probably as great or greater an area of racing and fine cruising than in all the British Isles. In Californian waters alone is cruising limited. There there are a few fine harbours filled to overflowing, but between them long stretches of unbroken coastline. In the Pacific Northwest, however, excellent cruising grounds again abound, and naturally it is in this area that yachting is growing at as rapid a rate as anywhere else. . . .

'In short, yachting in America is now active in an area extending from the latitude of the English Channel to that of half-way down the Red Sea or through the middle of the Sahara Desert. From west to east it covers an area comparable to that from Portugal to the Caspian Sea. Climatic conditions during the summer months favour the whole area while in the southern regions the sport flourishes all the year round. Small wonder that yachting now ranks as the most popular participant sport in the United States.'

To enumerate and detail the strengths of all the classes of small one-design racing yachts, larger than dinghies and with ballast keels, which have come into being or grown vastly in strength since 1945, would be impossible within a reasonable space, and perhaps tedious. One class, perhaps the most famous, and believed to be the largest keel boat class in the world, will be enough to reveal the nature of the new and universal taste for sailing, repeated yet more multitudinously in the smaller dinghy classes, which have spread like a contagion over the world.

The Dragon class was formed in Sweden as early as 1929 with practical and

The British 12-Metre *Flica II*, designed and built shortly before World War II. The style of crewing, with the crew sprawling over the topsides, belongs to the post-war period.

modest aims. It was intended to provide a cheap kind of small cruiser-racer with an exciting performance and the rugged accommodation of two berths beneath a cabin top – a boat for young people. The class spread throughout Scandinavia and along the Baltic coast, reached Britain via the Clyde in 1935, spreading farther to Ireland and the south coast.

After the war the rig was rejuvenated and various changes in the class rules made; and then the metamorphosis of the Dragon began. From being the young peoples' cheap knockabout cruiser-racer it evolved into a highly developed racing machine constructed by the most expensive yacht builders to the limit of the rule's tolerances and raced – even the sparse cabin became no more than a sail locker – by the most experienced helmsmen, many of whom were not young. Quaint illogicities appeared in this class, transformed from hack to high-bred racer. In one of the smallest, as in the biggest racing classes, perfection was sought with avid determination, regardless of cost. The rules originally, for reasons of economy, specified solid wooden masts. Pursuit of the highest possible performance, and hence, amongst much else, the lightest possible mast, led to the production of solid masts made from many pieces of timber which became more expensive to produce than the hollow masts which were prescribed in the cause of cheapness! Almost inevitably the class became one of those used for Olympiad yachting, and the Dragon, changed in almost everything but the basic hull form – by the standards of later days not an outstandingly good one – spread round the seacoasts of the world. Yet what occurred in Dragon class growth between the 1940s and the 1970s was repeated on a bigger scale in such smaller celebrated classes as the International Finn and International Flying Dutchman. The story of the Star class, which had begun it all, was being repeated now by numerous classes; explosive growth of numbers had become commonplace, the wonder had gone out of it.

By the early 1970s Dragons were being sailed in twenty-five countries and the total number of registered boats was about 1,755; and by the new standards of the period the boats no longer appeared small and were far from inexpensive. The five leading Dragon racing countries at this time were:

West Germany 312	Britain 294	U.S.A. 250
Sweden 172	Australia 141	

In proportion to their yachting populations it will be evident that the two last named countries were relatively the strongest supporters of the class. The distribution of Dragons was otherwise thus:

Switzerland 87	Spain 30	Philippines 14
Canada 84	Holland 30	Argentine 14
France 66	East Germany 28	United Arab Republic 9
Finland 36	Portugal 20	Bermuda 9
Austria 35	Norway 18	Lebanon 4
Belgium 33	Denmark 18	Uruguay 4
Italy 30	Hungary 17	

During the 1950s and 1960s the character of yachts was transformed by the use of glass-reinforced plastics and man-made fibres. In hardly more than a decade the world of the yachtsman underwent revolution. With each brisk year it became clearer that an ever higher proportion of yachts were going to be of the pedigree laboratory-out-of-factory. The galloping disappearance of timber disturbed many, and in the annual boat shows it was becoming ever harder to find craft whose hulls were planked in timber; until, by 1971, at the International Boat Show in London, among 117 yachts and motor sailers ranging in price between £450 and £30,000 there was one single wooden hull. At the first London show in 1955 the few plastics boats, all very small, had been regarded with curiosity. The commercial structure of the glass-reinforced plastics yacht industry encourages exhibition at boat shows, which tends slightly to over-emphasize the dominance of this form of construction; but at this period eighty per cent of the craft for which *Lloyd's Register* issued certificates were in plastic.

Each year the disappearance of natural materials from the structure of yachts became faster. No more hulls of timber which had grown to maturity by drinking sunlight where the summer trees stood like mountains of leaves. No more sails whose stuff had come from the hot, sad, glowing cotton fields where the Blues had been born. No more masts bursting open with a rose, for that matter; for masts had become just up-ended light aluminium alloy pipes, fit only to produce a plastic tulip. It all made for efficiency of course. Terylene sails and light alloy masts and spars produced boats with strength and sailing qualities unattainable when spars were of gold spruce and sails of the cotton duck which itself had seemed so revolutionary a few generations earlier. The ability of small modern yachts as seagoing and sailing machines is largely due to the use in their construction of these materials.

Also to the basic design of the craft themselves. The yacht had increasingly, during the 1950s and 1960s, been subject to the kind of scientific scrutiny that originated when the technique of tank testing them had been evolved. In the age of technicians the sailing problem came within the bounds of advanced aerodynamics, hydrodynamics and higher mathematics. This would have amazed our forefathers, who regarded sailing yachts more in the light of horses than of equations. The sailing yacht began to be described with chilling science as a combination of aerofoil and hydrofoil connected by a buoyant body. This view clearly discourages the aesthetic approach to yachts. There is indeed no evidence to suggest that the beauty of a yacht bears any relation to her effectiveness as a sailing machine. The old doctrine, so often applied to sailing yachts, and which guided so often the pencils of their creators at their drawing boards, that what looks right is right, over-emphasizes the power of aesthetic qualities on function; the newer doctrine, that what is right looks right, is a gross conceit of the scientist. Beauty and efficiency find themselves together only occasionally and by accident. But the well-known marine artist who described the aerodynamically efficient rigs of modern yachts as looking like a lot of damn dull isosceles triangles had his good reason.

By the 1950s the America's Cup, so deeply freighted with its load of history, seemed the epitome of yesterday's yachting, the ghost of a *flâneur* from the Grand Boulevards straying uneasily in a very public park on a bank holiday.

Yet, since 1946, there had been sporadic and tentative feelers put out in the cause of its revival in some form suitable for the modern world; but inevitably the tradition wreathed round the cup in the New York Yacht Club made this difficult. A course that might have seemed the most rational would have been to renew the contest using yachts of the finest type of the new day, the largest of the dual-purpose cruiser-racing yachts. Such might have been built to a high rating under the rule of the Cruising Club of America, or to the newer yet largely neglected Cruiser-Racer rule of the International Yacht Racing Union. Such a course would have produced a reversion to the kind of yacht that had

At the London International Boat Show in 1973, the centrepiece showing the pool and boats exhibited afloat. The first of the London shows was held in 1955.

originally been involved in America's Cup racing – *Genesta*, *Puritan*, *Mayflower*, *Galatea* – the finest yachts of their day with seagoing capability. But since then cup boats had become specialized creations beyond the range of ordinary yacht racing. So, it was decided, they must remain.

The pre-war 12-Metre class was selected, undignified in size compared with any previous America's Cup craft, though too large and specialized to have found any independent role in the yachting world of the 1950s. The America's Cup brought the Twelves a role more esoteric than any they could have aspired to during their many seasons of life prior to 1939, and any they could otherwise have aspired to after 1939. The selection of the class led to a further important change in the conditions governing the contest; that the challenger should cross the Atlantic on her own bottom. This was rescinded. It is a measure of the social importance of the cup that the changes in the Deed of Gift had to be approved by the Supreme Court of New York State.

A challenge on behalf of the Royal Yacht Squadron was issued in June 1957 and a syndicate of no less than a dozen members was formed to build a new 12-Metre. British preparations for the contest were conducted with what seemed great thoroughness. Four leading designers produced eight designs which were tank tested using the new facilities available in Britain at Saunders Roe in Cowes. But here already the seeds of failure were being sown, for this tank could not have the great experience of the tank-testing technique available at the Stevens tank in Hoboken, and lacked the yardsticks against which to assess performance. The selection fell upon an unusual design by David Boyd, whose success in the 6-Metre class with the pre-war *Circe* and several post-war Sixes could give confidence in his familiarity with the measurement rule. The new boat came out under the name *Sceptre*. Unlike any previous Twelve she had a large cockpit from which the crew operated, converting her into the resemblance of an open boat. A British designer was showing an ability in skinning the rules which rivalled that of the Americans in the past. But more seeds of failure were being strewn. There was no other up-to-date 12-Metre in British waters, though the pre-war *Evaine* was refitted to provide a rival for *Sceptre*. It was disconcerting when the new 12-Metre proved little if at all superior to *Evaine*, which was known to have been beaten so resoundingly by the American *Vim* in 1939.

And the Americans had not only *Vim* but three new boats built as contenders for the defence. Here was the familiar pattern of events: the elaboration and extravagance of the defence and, in comparison, the economy of the challenge. It was one boat against four. In a series of selection trials which provided the

finest 12-Metre racing ever yet known, the brilliant *Vim* almost proved herself worthy of being the defender; but eventually one of the three new boats, *Columbia*, like *Vim* designed by Olin Stephens, just contrived to gain the honour.

The contest itself lacked interest. The challenger lost four races in a row, never by less than seven minutes, once by more than eleven minutes. On the fourth race the challenger broke her boom; but such was the spirit of the helmsman and crew that the spar was fished and thus crippled she actually gained a little in the latter part of the race. It was widely agreed that the handling of the challenger was not inferior to that of the defender; but so great was the disparity of the boats that one of the most resounding defeats in America's Cup racing was sustained. This may have been partly due to the challenger's less good sails; the British were lagging behind in the development of sails in the new man-made fibres, as they had done so long ago in the making and setting of cotton sails. The big open cockpit, removing so much of the deck area, may also have produced aerodynamic effects lowering the sail plan's efficiency; in other respects the cockpit was a brilliant conception. Certainly the curious hull form of *Sceptre*, so bulbous in the bow and so fine in the stern, was wrong for the sea conditions in which the races were sailed.

Columbia, ahead, and *Sceptre* racing on a day of fresh wind in the 1958 America's Cup contest.

The first challenge for the America's Cup by Australia, in 1962, and the first also by any country except Britain since the second Canadian challenge in 1881, was an inspiring contrast. The Australians had chartered *Vim* against which to test their challenger, *Gretel*; and when the latter proved clearly superior to the gallant veteran, now approaching a quarter of a century in age, it was evident that a worthy challenger was to emerge. So she proved to be.

A new 12-Metre was built in America; but 1962 was one of the rare occasions when it was not a new boat that was selected for the defence. Instead *Weatherly* was chosen. One of the four Twelves built for the 1958 defence, she now proceeded to be selected against *Nefertiti*, the new boat, and *Columbia*, the yacht that had beaten her to the defence in 1958. The scene at Newport, Rhode Island as once more the America's Cup contest, the eighteenth, swung into action, has been described evocatively by Edwin P. Hoyt: 'Newport was busy that fall. Every hotel room was booked. Each crew had its own house, rented for several thousand dollars for the short season, and all the houses at Newport were in use, it seemed. Given the impetus of the America's Cup races, the old millionaires' playground was coming back into the public eye.'

What a strange affair indeed is America's Cup racing! Now, by 1962, so far removed from the vast and growing international world of yachting, sailing and boating, yet still a magnet to a very few hardened sailing experts, and still a beguiling circus to millions who are strangers to sailing affairs. Yet the contest, with its amazing power to survive in a transformed world, was itself – even unwillingly – inclining to modernity. The defender's crew this time was entirely amateur, and Hoyt has listed some of their professional occupations. On board *Weatherly* were found:

A real estate investor (B. Mosbacher, the helmsman)
A lithography salesman
A whisky salesman
A glass fibre salesman
An insurance broker
A photo studio operator
A shipping executive
Two owners of a firm of stevedores
Two college students

The fleet that turned out to watch the racing was believed to be the biggest in America's Cup history, and included the American President.

The Australians lost the first race, by 3 minutes 46 seconds, a smaller margin of defeat than in any race of the previous contest, but the second was won by 47 seconds. It was the first race to be gained by a challenger out of fourteen sailed since 1934; it was only the sixth race, out of fifty-four individual races, ever to be won by a challenger in the history of the Cup. *Gretel* lost the third race in light winds, which were considered unfavourable for her; but when she proceeded to lose the fourth race by a mere 26 seconds in winds even lighter, the worthiness and danger of the challenger were clear. There were some who considered *Gretel* the better boat; more who believed she was better handled. But she was beaten again in the fifth race by a time similiar to that of the first. It had proved to be one of the best challenges ever made for the Cup, though one race only was gained by the challenger. She had to join the lengthening list of failures.

Now a second British challenge was made, in 1964, and with results even more dispiriting than those of six years earlier, and sad indeed when compared with the forceful Australian challenge. David Boyd was again chosen to design the challenger, and the present author wrote of the new boat, *Sovereign*, at the time: 'It is most unusual for a yacht architect of long experience to change so completely his ideas on the shape he considers desirable for a racing yacht to a class rule as Boyd has done in producing *Sovereign* a few years after *Sceptre*. I doubt if an architect has ever produced two such dissimilar boats as *Sovereign* and

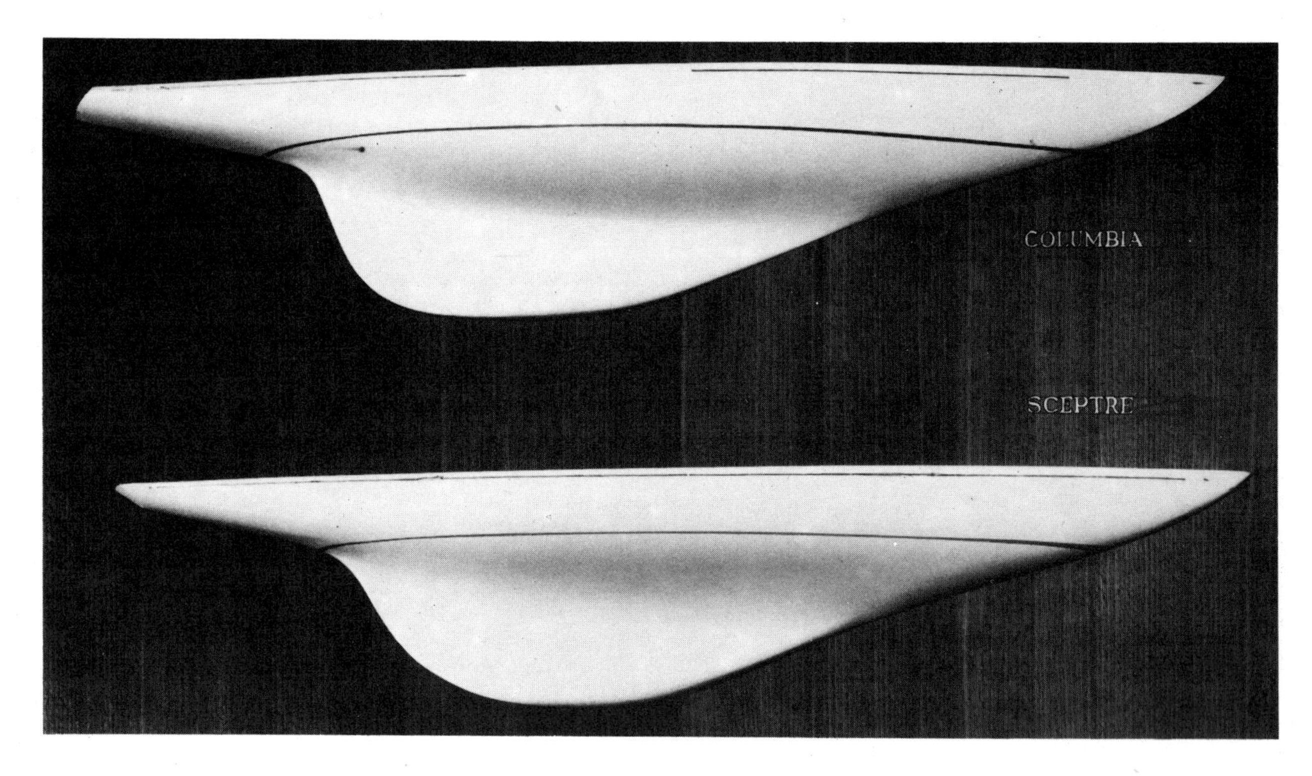

Sceptre to the same rule within an interval of a few years. . . .' But as in 1958 it had been disconcerting when *Evaine* frequently proved faster than *Sceptre*, now in 1963 people were even more uneasy when *Sceptre*, admittedly a much better tuned boat now, quite often defeated *Sovereign*. The radical change in shape had evidently not produced a radical change in performance.

During the 1963 season, when *Sovereign* came out a year before the challenge was to be made, she had, like *Sceptre*, no worthy 12-Metre against which to compete; and the need was not adequately supplied when the Australian brothers Frank and John Livingstone generously ordered a sister ship to *Sovereign* to provide a rival in 1964. The pressure of time made it impossible to have a new and competitive design produced. The new yacht was *Kurrewa V*. In America two new 12-Metres were produced to compete against the already formidable fleet of Twelves in existence on the western side of the Atlantic. One of these, *Constellation*, again by Olin Stephens, was selected. At the helm of *Sovereign* was Peter Scott, famous as an ornithologist and painter and one of the pioneers of dinghy racing during the pre-war days.

Like *Sceptre*, *Sovereign* lost four races one after the other, the second of them by the staggering time of 20 minutes 24 seconds, one of the most complete defeats in the history of America's Cup racing, so rich in inadequate challengers. It was a defeat of the kind that could only be acutely embarrassing for winner and loser alike. And that was the tone of the whole contest.

When Australia challenged again, in 1967, Olin Stephens, in *Intrepid*, produced a boat for the defence not perhaps quite so outstanding as *Ranger* and *Vim* had been in their days, but nevertheless an advanced yacht and unquestionably the best 12-Metre yet to go afloat. She was also the ugliest Twelve yet to have appeared, with snubbed weight-saving overhangs, and taking every advantage of the rule, even to the extent of making such furnishing as there was, poked into corners, of the very light and – for joinerwork – most unsuitable balsa wood.

Half-models at the Royal Thames Yacht Club, London. Opposite *Sceptre* and *Columbia*, the America's Cup defender in 1958, when *Sceptre* was severely outclassed. Below *Sovereign* and *Constellation*. *Sovereign* was the second post World War II challenger for the America's Cup. Though widely different in design from *Sceptre* she was no less inferior to the defender in the America's Cup series of 1964.

The ice-blue Australian *Dame Pattie* was beaten by the white American yacht,

which could work out to windward so finely, in four straight races. After the former Australian challenge it had been said that had the contest been sailed in stronger winds, of twenty to twenty-five knots, *Gretel* might have carried the cup back to Australia. *Dame Pattie* had been designed for best performance in winds under fifteen knots. It had not availed.

France entered America's Cup racing for the first time in 1970 when altered conditions for the contest enabled international elimination trials for the challenger to be held. While the conditions for the defence became ever tougher as the terms for the challenge were eased, the intensity of effort put forth by challengers increased. In Australia the training of crews had been in progress since 1968, with *Gretel*, the 1962 challenger, *Dame Pattie* the 1967 challenger and the old U.S. 12-Metre *Vim* engaged. The new Australian Twelve, *Gretel II*, though initially disappointing, was selected to race against the French Twelve *France*, herself chosen from a team comprising the British *Sovereign*, challenger in 1964 and *Kurrewa*, built for the same contest. *Gretel II* was selected as challenger in the trials with *France*. In the U.S.A. two new 12-Metres and the older *Weatherly*, defender in 1962, fought for selection with the previous defender *Intrepid*, which was selected. Such activity in challenger and defence, and the international character of the challenge brought a new dimension to America's Cup racing.

The racing between *Intrepid* and *Gretel II* made one of the closest contests in America's Cup history, though the challenger gained one race only. The second race resulted in the disqualification of *Gretel II* as a result of a collision at the start, and for the first time in cup history the challenger lost a race following a protest decision of a foul. It was not until the fourth race that the challenger gained a victory, and *Intrepid* proceeded to win the final race. The close matching of the two boats, however, made the results a matter of speculation throughout.

Opposite

An impression by Raoul Dufy in 1936, of the Platform at the Royal Yacht Squadron, Cowes.

14 Cruiser-racer

In the years following World War II the formerly small and peripheral activity of offshore racing began taking control of all yacht racing other than in the small class yachts without accommodation. Racing in bigger craft moved *en bloc* to the cruiser-racer.

In European waters the Royal Ocean Racing Club entered the post-war world armed with a rating rule, a system of time allowance, an organization and outstanding virility. The fact that the club was ready, as soon as yacht racing might be resumed, to take the lead, gave it a great advantage over the inevitably slower moving International Yacht Racing Union, which realized equally well – as indicated above – that the cruiser-racer must become the racing yacht of the near future.

The rule of the Royal Ocean Racing Club, as the formulation stood in the immediate post-war years, was well devised for its purpose; which was to measure any good type of yacht for racing offshore, and to do so while the yacht lay afloat and without overmuch complication. Thus was created the big fleet of yachts with Royal Ocean Racing Club ratings able to race under time allowance in events ranging from the Fastnet and Channel races to those of local club regattas.

During these years Commander John Illingworth, one of the most influential figures in international yachting during the two post-war decades, was for a time Commodore of both the Royal Ocean Racing Club and the Royal Naval Sailing Association. He was a natural organizer, and an outstanding offshore skipper and owner. His mind had the bent of the trained engineer with the addition of a flair for creative design. His attitudes were international, never provincial. His were an unusual combination of qualities, carried with a modesty that rarely failed to charm.

When, in 1946, he and his architect, J. Laurent Giles, produced the *Myth of Malham*, a yacht appeared on the waters as impressive, and subsequently influential, as the *America* in her day, the *Gloriana*, the *Britannia* and her contemporaries, or the J-class *Ranger* and 12-Metre *Vim*. It is not an exaggeration to suggest that a new era was born when the *Myth of Malham* took shape amongst the Scottish winter snows of 1947, in the yard of H. McLean & Son, Gourock. The boat, with her snubbed overhangs, high freeboard, straight sheerline and light displacement – all judged by the standards of the time – represented a successful attack upon the fragilities of the Royal Ocean Racing Club's measurement rule. An American authority, faced with *Myth of Malham*, suggested that the rule be hove overboard before ocean racing became contaminated by more *Myths*. Yet she prefigured the future, announcing with some stridency the arrival of a new type of yacht for racing offshore. Her outstanding features were a direct outcome of evading the system of measurement. The high freeboard and straight sheerline gave her big internal depth but relatively little hull in the water except for the deep fin-like keel; the snubbed overhangs allowed the maximum

Motor-sailer *Blue Leopard*, one of the most advanced yachts of her type, of light displacement and high power engines, with high performance under both sail and power. Painting by David Cobb.

effective sailing length for a small measured length. *Myth of Malham* was one more in the long line of rule cheaters. But her superb performance – she won the Fastnet and Channel races in 1947 – demonstrated merits of the light displacement type of yacht not hitherto accepted as suitable for racing offshore. Her designer, Laurent Giles, wrote at this time: 'When John Illingworth commissioned my firm to design his new ocean racer in 1946 I took good care to plant the responsibility for her behaviour at sea firmly on his shoulders . . . *Myth of Malham* justified him amply and ably. She was absolutely at home on the high seas.'

Myth of Malham, with snubbed ends and shallow body, high freeboard and no curve in the sheerline, set the new course in design for post-war ocean racing when she came out in 1947. She takes her place among the few yachts, such as *Gloriana* (see pp. 77 and 79) and *Jullanar* (see p. 59) which have exerted a permanent influence on yachting.

At the same time Colonel H. G. Hasler was reinforcing the claims of light displacement by racing offshore one of the very light 30 Square Metre class of Scandinavian inshore racers, the *Tre Sang*, a yacht the antithesis of what had formerly been considered fit for deep water work. The age of the light displacement ocean-going yachts had opened. The proof of their quality followed by their wide acceptance was the most notable feature differentiating offshore racing in the post-war era from any earlier time.

There was another difference. Yachts rapidly became not only lighter in relation to their length but generally smaller. Everything became diminished except numbers. In 1947 the minimum waterline length for the Fastnet race was lowered from 35 ft to 30 ft; it was shortly reduced again and in 1953 the Fastnet Cup was won by Sir Michael Newton's *Favona*, which was of the minimum waterline length of 24 ft. The combined effects of smaller yachts and yachts lighter in proportion to their size produced that characteristic breed of the post-war yachting world, which earlier periods of yachting would have regarded as yachts in miniature. The source of particular amazement to the earlier generation would have been the seagoing ability of these miniature creations. In 1950 a transatlantic race was held for which the minimum waterline length was 24 ft, and the years were opening when the North Atlantic in summer was to become almost a highway for craft of this size and smaller. John Illingworth once drew attention to the fact that three of the five boats in the 1950 Transatlantic race had less waterline length than the breadth of the schooner *Atlantic*, which won the Transatlantic race in 1905.

It was in this spirit of the age that the Junior Offshore Group was founded in Britain in 1950 and the Midget Ocean Racing Club in the U.S.A. The object of the Junior Offshore Group on its formation in 1950 was to provide offshore racing for boats of between 16 ft and 20 ft. Though it seemed an adventurous organization rich in the unusual when it was founded, not a little owing to the example of Patrick Ellam in his canoe *Theta*, which he sailed far offshore, the Junior Offshore Group classes, using the sail emblem of theta, are now an established institution, and their boats no longer seem small. This is the measure of the times.

By 1965, the thirtieth year of ocean racing in European waters, the fleet of yachts measured by the Royal Ocean Racing Club's rule was almost one thousand strong, and the club was responsible for the measurement of more yachts outside its own membership than within it. This number did not include those measured independently in Holland, Germany, France, Scandinavia, Italy, Spain and Australia, which amounted to about five hundred. The figures indicate the role that the club had now assumed as an international organization reaching far beyond its status of a private club. The club had certainly not been born great in 1925; whether its later position was the outcome of its achieving greatness or having greatness thrust upon it would make a nice study. But there was no question that in the body politic of international yachting it now performed functions properly belonging to the International Yacht Racing Union. It was a curious development.

In numbers of yachts involved the 1965 offshore racing season established a new record. Out of a thousand entries for fourteen races 965 yachts actually started. The secretary, Alan Paul, in his account of the season wrote: 'So ended a wonderful season of offshore racing in which more yachts and more people took part than ever before. Undoubtedly there have been improvements in hull design in recent years but far greater advances have been made in the standards of sail and gear and in the competence with which these were handled. The

Admiral's Cup and One Ton Cup contests have injected a new, highly competitive spirit into ocean racing, which some of the older boats find difficult to maintain.'

The last sentence should particularly be noticed. That the older boats, which were of the kind for which ocean racing had been established, were falling behind in the competition was a measure of the dedication to pure racing that had arisen; and of this the two new contests mentioned above were the expression.

In 1957 a new international yacht racing contest was organized which since then has grown steadily in repute. It has been described, with exaggeration, as comparable in its status to the Ashes or Wimbledon, and more accurately as worthier than the America's Cup.

Cowes Parade and the fleet of cruiser-racers just offshore at the start of the Queen's Cup race during Cowes Week 1965.

The Admiral's Cup was presented by five members of the Royal Ocean Racing Club, Sir Myles Wyatt (Admiral of the club in 1957), Captain John Illingworth (Commodore 1948–50), Peter Green (Commodore 1961–4), Geoffrey Pattinson (Rear Commodore 1960–2) and Selwyn Slater. An important object of the cup was to draw foreign yachtsmen to British waters in alternate years for the Fastnet race and Cowes Week. It is thus always to be held in the U.K.

The organization of the event was an original development of the already well-known system of team racing. National teams were to be composed originally of up to three boats; since then the maximum of three boats has become obligatory. They were to take part, however, in a series of both offshore and inshore races. Originally these were the two chief events in the European offshore racing season, the Channel race and Fastnet, and the two most prominent races of Cowes Week, the Britannia Cup and the New York Yacht

Club Cup. This mixture of two different kinds of race, together with the fact that the yachts involved might vary widely in size and type and raced under time allowance, introduced a new form of racing competition into yachting, which the experience of nine successive contests has proved to be admirable. Results are based on a points system, points being received by every boat completing a race in relation to her position on corrected time; they are biased heavily towards the two offshore races, and particularly towards the Fastnet race, which earns three times the number of points of the inshore races.

The story of the Admiral's Cup so far is summarized in the table.

Year	Argentina	Australia	Belgium	Bermuda	Brazil	Britain	Denmark	Finland	France	Germany	Holland	Ireland	Italy	Portugal	U.S.A.	South Africa	Spain	Sweden	Total Teams	Winners
1957						•									•				2	Britain
1959						•			•		•								3	Britain
1961						•			•		•				•			•	5	U.S.A.
1963						•			•	•	•				•			•	6	Britain
1965		•				•			•	•	•	•			•			•	8	Britain
1967		•				•		•	•	•	•	•			•		•		9	Australia
1969	•	•		•		•		•	•	•	•		•		•		•		11	U.S.A.
1971	•	•	•	•	•	•			•	•	•	•	•		•	•		•	14	Britain
1973	•	•	•	•	•	•	•	•	•	•	•	•	•	•	•	•			16	Germany

Note: *In 1971 one-yacht teams were allowed from Austria and New Zealand. One-yacht teams were not allowed in 1973.*

The undeviating steadiness of the growth in numbers and the widening spread geographically of competing nations is impressive. Beginning modestly and attracting in 1957 only two nations, Britain and the U.S.A., the biennial events have grown steadily, with three nations concerned in 1959, five in 1961, six in 1963, eight in 1965, nine in 1967. The contest in 1969 raised eleven nations, in 1971 fourteen, and the growth continued, to reach sixteen in 1973. It may reasonably be argued that the Admiral's Cup now contributes more to the healthy development of yachting than the America's Cup, though the public interest it can arouse is small in comparison. The yachting correspondent of a London newspaper was heard to complain: 'It's all very well my office giving me what they describe as a generous eight hundred words, but I need seven hundred just to describe how the affair is organized.' The complex system of points, the fact that time allowance is involved and possibly as many as between forty and fifty boats in four races, two of them far offshore, precludes wide public interest.

The first contest in 1957 was a private kind of affair with little official organization. There was a Transatlantic race from Newport, U.S.A., to Santander, Spain, that year, which included Swedish and German entries and three American yachts. The Americans came in successively first, second and third, the winner being *Carina*, which made the record of being the first yacht to win two Transatlantic races. She proceeded to win one of the toughest of Fastnets; but the British team of *Jocasta*, *Myth of Malham* and *Uomie*, owned by three of those who had presented the Admiral's Cup, was able to defeat the American. The U.S.A. was not represented in the next contest, which again was won by Britain.

In the 1961 races the U.S.A. won, thanks mainly to their fine performance in the offshore races. By 1963 international interest in Admiral's Cup racing had grown, while in Britain for the first time there were selection trials, with six yachts competing for the three team places. Once again the American team came over in a transatlantic racing fleet. Fortunes swung violently in the Admiral's Cup series, with Sweden gaining some thirty points lead in the Channel race. The Swedish yacht *Staika III* proceeded to win the Britannia Cup in Cowes Week, and by this time Britain lay third in the contest, led by Sweden and the U.S.A. But the British team took the first three places in the New York Yacht Club Cup, and a member of the team won the Fastnet. This allowed Britain to recapture the cup from the U.S.A. Again in the following year, against eight nations, Britain won. For the first time Australia entered a team, and proceeded to give a demonstration of the sailing flair and relentless organization that had been revealed in Australia's first challenge for the America's Cup three years earlier. After Britain's thinnest of leads – two points only – over Australia following the Channel race, and in spite of Australia's *Caprice of Huon*, a

American offshore racing yacht, *Carina*.

thirteen-year-old yacht, winning both the Britannia and New York Yacht Club Cups, Britain had increased her lead, though still only to a slender fourteen points, by the end of Cowes Week. But Britain collected enough points in the Fastnet race to carry off the cup, her fourth win in the five contests up to this time.

Australia's worthy victory came in 1967, when twenty-two boats representing nine nations assembled. In light winds sixty yachts raced round the Britannia Cup course, the Admiral's Cup teams amongst them, to produce a win for Australia, with British team boats in the second and third places. The New

Caprice of Huon, English designed and built in Tasmania in 1951, was thirteen years old when she won the Britannia and New York Yacht Club cups at Cowes in 1963 as a member of the Australian Admiral's Cup team.

A rush downwind in the Sydney–Hobart race.

York Yacht Club Cup also went to an Australian boat, the magnificent *Rabbit II*. In the Fastnet race, won by *Pen Duick III* of the French team, the Australians gaining second and third places overall in the large fleet, increased their already commanding lead in points to produce an outstanding Admiral's Cup victory: Australia 495 points with Great Britain following with 391 points, the United States third with 358.

Australia approached the defence of the cup in 1969 with a thoroughness that tended, in Australian waters that season, to overshadow the preparations for their challenge for the America's Cup due to take place in 1971. With *Ragamuffin*, *Koomooloo* and *Mercedes III* the Australians produced for their team two new boats built specially for the contest and one that had sailed in the victory of 1967. In the first offshore event of the series, the Channel race, and in the second Cowes inshore race, the New York Yacht Club Cup, Australia gained a commanding number of points over the eleven other teams, and she was second to Britain in the Britannia Cup. Australia's position appeared safe enough before the Fastnet, with 317 points comfortably leading Britain's 273 points, followed by Italy with 250 and the U.S.A. with 244 points.

The Fastnet was a slow, light wind affair ruled to an unusual extent by chance, equally unsatisfactory for the thirty-one yachts of the Admiral's Cup teams and the 150 other competitors, and fatal in its whims to the well-placed Australian team. Struggling in light airs and calms, the U.S.A. team was able to add enough points to its score to overtake the leaders. The final standings were U.S.A. with 496 points followed by Australia 482 points; Britain 471 points; Italy 451 points.

The growing intensity of Admiral's Cup competition led to a change in the rules governing the 1971 series. The density of the yacht population during Cowes Week has produced what has been called jolly yachting obstacle races. It was considered undesirable to have the fiercely contending Admiral's Cup boats embroiled in the turmoil of all the rest during the Britannia and New York Yacht Club races; circumstances that led to unfortunate incidents. Conditions differed in the two offshore races when the yachts, once the starting line dropped astern, were not sailing at close quarters. It was therefore arranged that the Admiral's Cup fleet should race on its own in the two races of the Week counting for points, a measure of exclusiveness indicating the growing importance of the affair.

The preparations for the contest indicated this no less. For the British team in 1971 twenty-seven first-class yachts competed for the three places in the team, none of them older than the 1968 commissioning, and sixteen of them being new that year. The contenders were said to have been worth half a million pounds. Amongst them was the British Prime Minister's yacht, the second *Morning Cloud.* No former Prime Minister had ever been an expert yachtsman; one of them, the Victorian Lord Rosebery, had confessed to hating Cowes. In this year the Australian team dropped to third place. The British team, captained by the Prime Minister who had been selected to the position on the exceptional merits of his boat and his organizing ability, despite the fears that other affairs might prove a distraction, led the British team to its fifth victory; the U.S.A. was third. An Austrian entry appeared this year, representative of one of the few nations in Europe still mainly unaware of yachting.

By 1965 there had passed twenty years since the war during which racing in the larger yachts had been wholly – excepting three challenges for the America's Cup involving a couple of handfuls of 12-Metres – in cruiser-racers governed by the highly developed time allowance systems of the Royal Ocean Racing Club and Cruising Club of America. There appeared the inevitable tendency for the cruiser-racers in the premier flight to become more racer than cruiser. The old process of 'men against the rule' was operating again, the earnest search for the best possible racing machine under the governing terms of measurement. The shape of the dual purpose cruiser-racer was being refined and pinched into the

Rabbit, of the One Ton Cup class, to fixed rating under the International Offshore rule.

racing yacht which rarely cruised, while the earnest thoughts of economy that had governed the earlier post-war years were losing their urgency. Pure racing yachts, able to discard the unavoidable crudities of time allowance were being demanded again, craft that could replace the now extinguished thoroughbreds of the pre-war 6-Metres and 8-Metres.

In 1965 the famous old racing trophy of the defunct 6-Metre class was offered again for competition by the *Cercle de la Voile de Paris*. The Commodore of this body, Jean Peytel, after consultation with John Illingworth, offered the One Ton Cup for racing between yachts rating at 22 ft under the Royal Ocean Racing Club rule. Forty years after the club had been established to encourage ocean racing between seagoing yachts of various types and sizes competing under time allowance, the club set about fostering the purest type of racing yachts of fixed rating competing on level terms without time allowance. There were those in the club who felt their birthright was in hazard. Such a development was inevitable so long as yacht racing remained prosperous and encouraged the pursuit of the highest performance possible. It might not have been expected, however, that the controlling body should be the Royal Ocean Racing Club; there was the International Yacht Racing Union with its fixed rating Cruiser-Racer classes, which might have seemed the obvious choice for the new purpose. It was a measure of their failure and the success of the Royal Ocean Racing Club's types of yacht that the club should have in effect taken over an important section of pure class racing.

The One Ton Cup contest was to comprise a series of three races, two over inshore courses and one offshore. It will be evident that the emphasis, unlike that of the Admiral's Cup contest, was on regatta sailing. With yachts being produced to fixed rating under the Ocean Racing rule – a relatively flexible compilation not originally intended for the purpose – the art of design was put into a hotter crucible than it had known for many years. In a short time a new class of very expensive highly refined racing machines was begotten, following the first of the revived One Ton Cup racing off Le Havre in 1965. Some time later a commentator wrote: 'All will agree that the One Ton Cup has stolen more headlines, built more boats, generated more hard thoughts than any other events except the America's and Admiral's Cups in the last decade.'

The revival of racing between yachts of fixed rating but free design proceeded to spread. The lean years of the 1940s and 1950s were becoming a memory. A year after the re-establishment of the One Ton Cup, Half Ton class racing was established, for yachts of 18 ft rating. In the Mediterranean the preference for larger yachts was demonstrated when the Yacht Club Italiano organized a first series of races for a Two Ton class, for yachts rating at 28 ft. It will be noticed that the 'Ton Cup' designation has been adapted for the various sizes. The term 'Ton' in this connection is of purely historical derivation, now unrelated to tonnage in any other sense; it was a pleasing memorial to the fact that the original Ton Cup was once raced for by yachts measuring one ton under the nineteenth-century measurement rule.

The international organization of yachting underwent a crucial change in 1969, as the result of an agreement between the Cruising Club of America and the Royal Ocean Racing Club, to adopt a common international rule of measurement for offshore racing yachts. Since the use of the rules had become world wide, being used for short distance handicap racing as well as that offshore, the effect of the change was far-reaching. On the merits of a common rule for the world there were wide differences of opinion, but of the remarkable position now attained by these two clubs there could be no question. A little wryly some now recalled how prominent early members of the Cruising Club of America nearly fifty years earlier had objected to the club sponsoring racing; and, a little surprised, some remembered what a small and eccentric organization on the fringes of yachting had the Ocean Racing Club been in the 1920s.

A new rule, subsequently to become known as the International Offshore Rule, was formulated by an international technical committee, most of whose members belong to the Royal Ocean Racing Club and all to this or the Royal Cruising Club of America. It was headed by Olin Stephens, the American yacht architect who now, some forty years after he had sprung to notice with the yacht *Dorade*, had assumed the mantle of Nathaniel Herreshoff, and had no rivals for the role of the world's leading yacht designer. The rule that emerged from the Committee's deliberations would have staggered yachtsmen prior to 1914; it would have surprised most of them prior to 1939; it made uneasy those in 1969. A creation of massive complexity and of mathematics that was elaborate but not, in the opinion of some capable of judging, of much elegance, it called in the computers of the new age to aid the processes of yacht design. The pursuit of perfection under the spur of racing was producing jarring effects. 'What art thou thinking there below' some people were inclined to inquire of the founding fathers of offshore racing.

The emergence of a single international rule governing the design of all the major types of racing yacht throughout the world was a step in the history of yachting whose sweeping implications cannot yet be focused. We live in the dawn of this new yachting era.

In September 1973 a fleet of yachts seventeen strong was started on a race round the world from Henry VIIIth's castle in Southsea, off the entrance of Portsmouth harbour. The yachts were of a size impressive to the eyes of 1973 and they were setting off on a new kind of yachting contest, related on the one hand to the ocean marathons of the singlehanders and on the other to conventional ocean racing.

Essentially the contest was composed of four ocean races, each in a grander manner than any formerly envisaged. The circumnavigation was to be completed in four stages, each of them a separate ocean race: Portsmouth to Captetown, 6,900 miles; Capetown to Sydney, 6,500 miles; Sydney to Rio de Janeiro, 8,370 miles; Rio to Portsmouth, 5,560 miles. The classic ocean races which had by this time become established, the Newport–Bermuda, the Fastnet, Sydney–Hobart and the more recent Mediterranean Middle Sea race range in length between 600 and 635 miles. The longest conventional ocean races, the Buenos Aires to Rio de Janeiro, held eight times since 1947, is 1,200 miles. These figures indicate how grand indeed were the four races comprising the circumnaviation, the longest leg of which, from Sydney to Rio de Janeiro, was about seven times the length of the above Buenos Aires–Rio race and about fourteen times that of the Fastnet. When the yachts left Portsmouth bound for Capetown they were starting a race equivalent to ten times round the Fastnet. Yachts entered for any one or more of the four legs, and the overall winner was the one gaining the greatest aggregate points in the four races combined. The yachts, some of those that started from Portsmouth, others that joined on one of the later legs of the course, arrived back in Portsmouth during April 1974, the Mexican ketch *Sayula II* being the winner.

While much, apart from the distances involved, follows the offshore racing practices that have become familiar, this race inaugurated a new role for large ocean-going yachts, a number of which were designed especially for the purpose.

In the Fastnet race of 1973 the biggest boats, grouped in Class 1, rated at 33 ft to 70 ft under the International Offshore Rule. Only yachts in this size range are eligible for the Whitbread contest. There were five classes, and a separate class for the Admiral's Cup yachts, in the Fastnet, and only some eighty of nearly three hundred yachts entered, were above 33 ft rating. In the whole fleet there were only three boats exceeding 50 ft rating. For the Whitbread races a new breed of large ocean-racing yachts came into existence, a welcome deviation from the prevalent form in the age of midgets. The steady trend towards ever smaller yachts has persisted since 1945 – indeed since 1919 when it began – and it is appropriate that the organization of yachting should produce a reaction, however small, against it.

Looking into the future there are signs that this may persist. While preparations were being made for the Whitbread race notices were appearing of a further racing circumnavigation in 1975–6. Sponsored by the London Financial Times, under the supervision of the Royal Ocean Racing Club, and entitled the Financial Times Clipper Race, the course proposed is one to enable the progress of ocean yachts and yachting to be historically evaluated. The yachts will sail the clipper ships' course from London to Sydney round the Cape of Good Hope, returning from Sydney round Cape Horn – the ocean passages made by the wool clippers in the 'seventies of the last century, when *Cutty Sark* completed them in 72 and 73 days respectively and her rival *Thermopylae* in 75 and 77 days.

15
A world of dinghies

To appreciate at its full significance the world of dinghy sailing that has sprung into existence since World War II it is desirable to return briefly to the inter-war period.

'Dinghies are the future' was a not uncommon saying even then amongst those who prided themselves on farsightedness. But the estuaries, rivers and reservoirs were not dense with dinghies during those years. Small boat sailing was of no great stature within the sport. In 1939 the most advanced of the dinghy world was represented in Britain by the International 14 ft class and the National 12 ft class. The former had brought prestige to dinghy racing and a certain chic; yet in 1939 the class had been international for only twelve years; only for twelve years had there been a Prince of Wales Cup; and how appropriate it was that the Prince of Wales should have presented it! Here is a rich touch of period atmosphere, of the days when the girl was so thrilled because she had danced with a

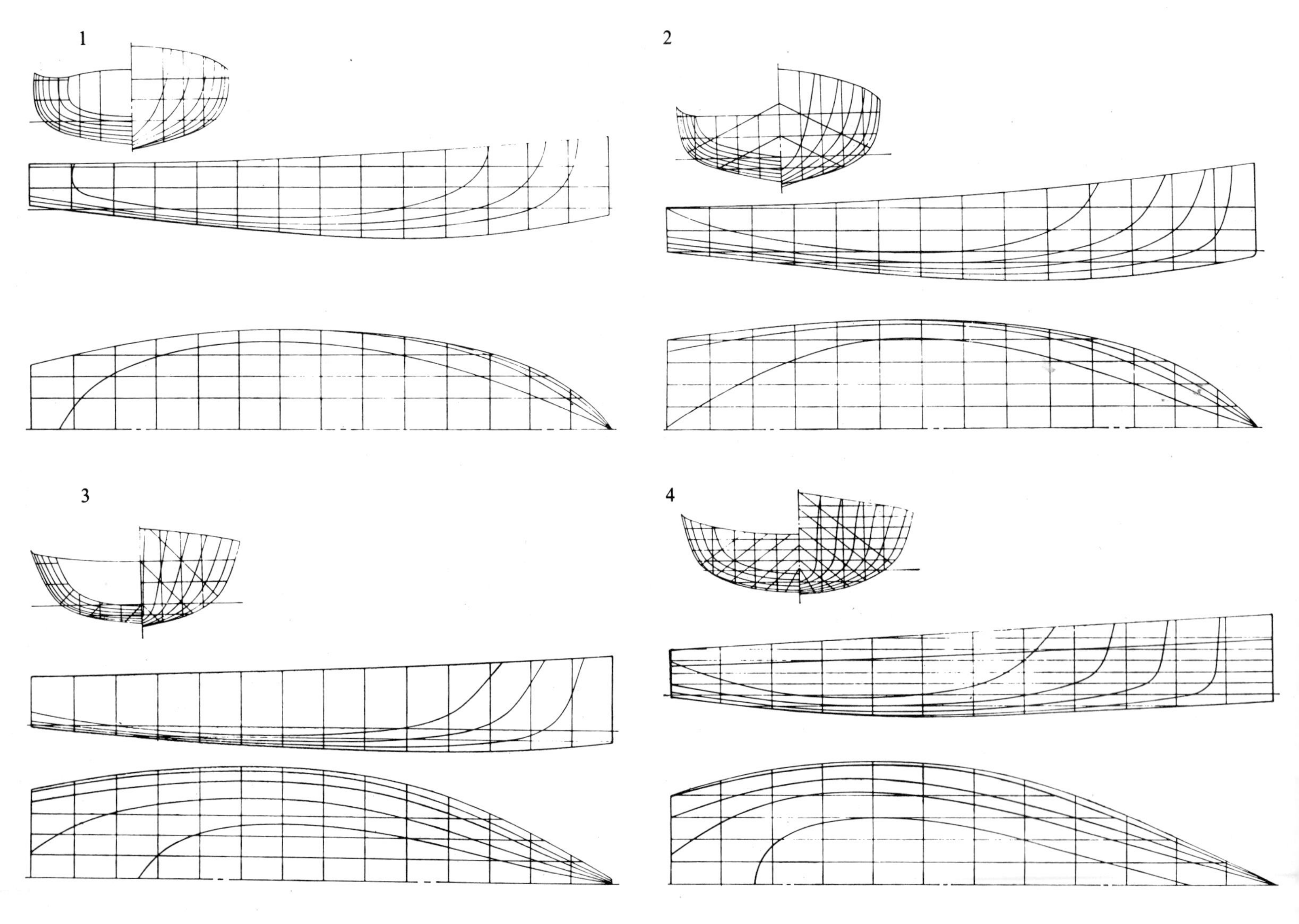

Opposite

Changes in design in the International 14 ft class between 1935 and 1971. The various forms adopted under the changing influences show the creative interest released in a class where design is not stereotyped.

1 *Alarm*, 1935, designed by Uffa Fox for Stewart Morris. With a relatively narrow transom and deep rocker, she won three P.O.W. Cup races.

2 *Windsprite*, Austin Farrar's 1950 design for Bruce Banks won the P.O.W. Cup four times. A wider, shallower transom with smooth graceful lines.

3 *Volcano*, Bruce Kirby's 1969 Mark V design, was one of the first hulls designed for the extra power of trapezes. She has very little rocker and flatter sections for greater stability.

4 *Windhustler*, the 1971 Cracknell Mark V, with the finest bow above the waterline and fattest U-sections below the water, ever seen in the class.

Below

Snipe class boats rounding a mark at Larchmont in 1934. This simplest of boats, 15 ft 6 in long, was designed in 1931 and sponsored by the American journal *The Rudder*. Thirty years later there were some 15,000 *Snipes* in twenty-eight countries.

chap who had danced with a girl who had danced with the Prince of Wales.

The International 14 ft class has claims to be the forerunner and inspiration of modern dinghy racing. The origin of this class is to be found as early as 1901 with the rules for what were then known as the West of England Conference dinghies, which stipulated boats having an overall length of 14 ft, draught not to exceed one foot, no bulb on the centreboard and no outside ballast. There was a further requirement that the dinghies were to be 'unsinkable'. The rules strike a future note. The West of England boats provided a model on which to base the National 14 ft class, which became International in 1927, the year when the first Prince of Wales Cup was sailed by a fleet of forty-one boats. Subject to the rule limitations, design in the class was free and soon advanced rapidly. Some of the leading characteristics of the multitudinous types of dinghies yet in the future were incubated in the pioneering 14-ft Nationals and Internationals. They further instituted an idea then new. There were already a number of different dinghy classes in existence, indigenous to their own part of the coast, and so limited to local racing. With the Fourteens there appeared the possibility of a nation-wide class, and then one with an international spread, bringing together at open meetings ideas on design and skills in handling. This was a major step towards the modern world of dinghy racing.

It was not until as late as 1936 that the National 12 ft class was formed by the Yacht Racing Association, and it was thus a mere three years old when war came. Essentially it was a cheaper version of the International 14 ft class, clinker built but like the former of open design capable of development. The two classes,

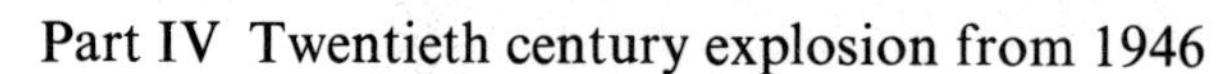

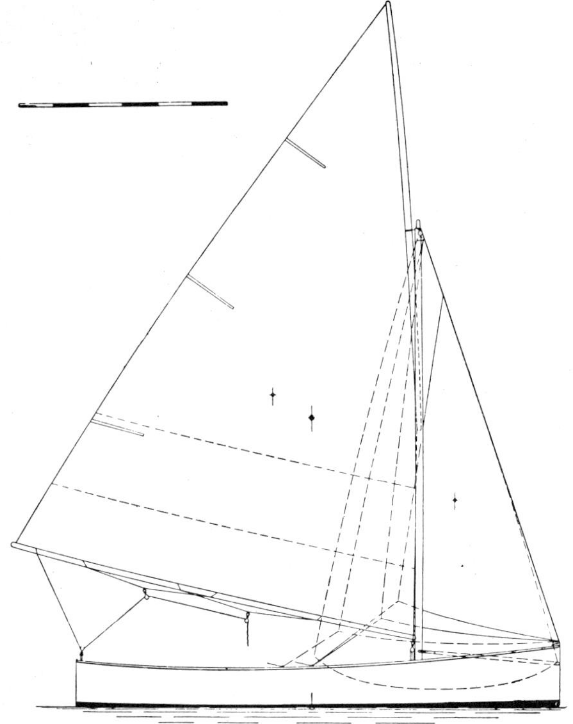

Snark (above and below) represents a first-class 14 ft racing dinghy of 1911. She was an undecked craft with a hull shape that would not plane readily. The gunter-lug sail plan shows the typical dinghy practice of the period.

Opposite

A Flying Fifteen.

and they alone, anticipated the future of dinghy racing. Yet the average age of the two classes in 1939 was a mere eight years. By 1951 the 12-ft Nationals formed the biggest dinghy class in the world that was not one-design; there were then 1,000 sail numbers allocated, but unlike the International 14 ft class there were few boats outside Britain.

There was nothing comparable to these two classes in the U.S.A. Small boat sailing on a large scale was based in the Star class and subsequently on the smaller Snipe. The latter class of 16 ft half-decked chine hulls was founded in 1931 at a meeting of the Florida West Coast Racing Association. The design was sponsored by the magazine *Rudder*, and provides an early example of the launching by sponsorship of a dinghy class that, thanks to the publicity, attains striking popularity. By the mid 1950s there had been issued 11,000 sail numbers to the class. This was a markedly different style of development to that of the 14-ft Internationals, or even the 12-ft Nationals. The Snipe were cheap boats with their chine hull adapted for amateur construction. They were simple and of one design, hence incapable of technical progress, and in the cause of cheapness and simplicity their minimum weight was limited to 425 lb or more than double that of more sophisticated dinghies. The 14-ft Internationals, in contrast to the Snipe, evolved into what was once described as 'the most royal of dinghy classes with almost no limits set on the improvements an owner may make'. They set 145 sq ft of sail compared with the Snipe's 100 sq ft, and the wooden carvel planked boats were the most superb examples of the professional boatbuilder's art ever to have appeared. In spite of their expense the boats were soon to be found in Canada, New Zealand, Bermuda and the United States.

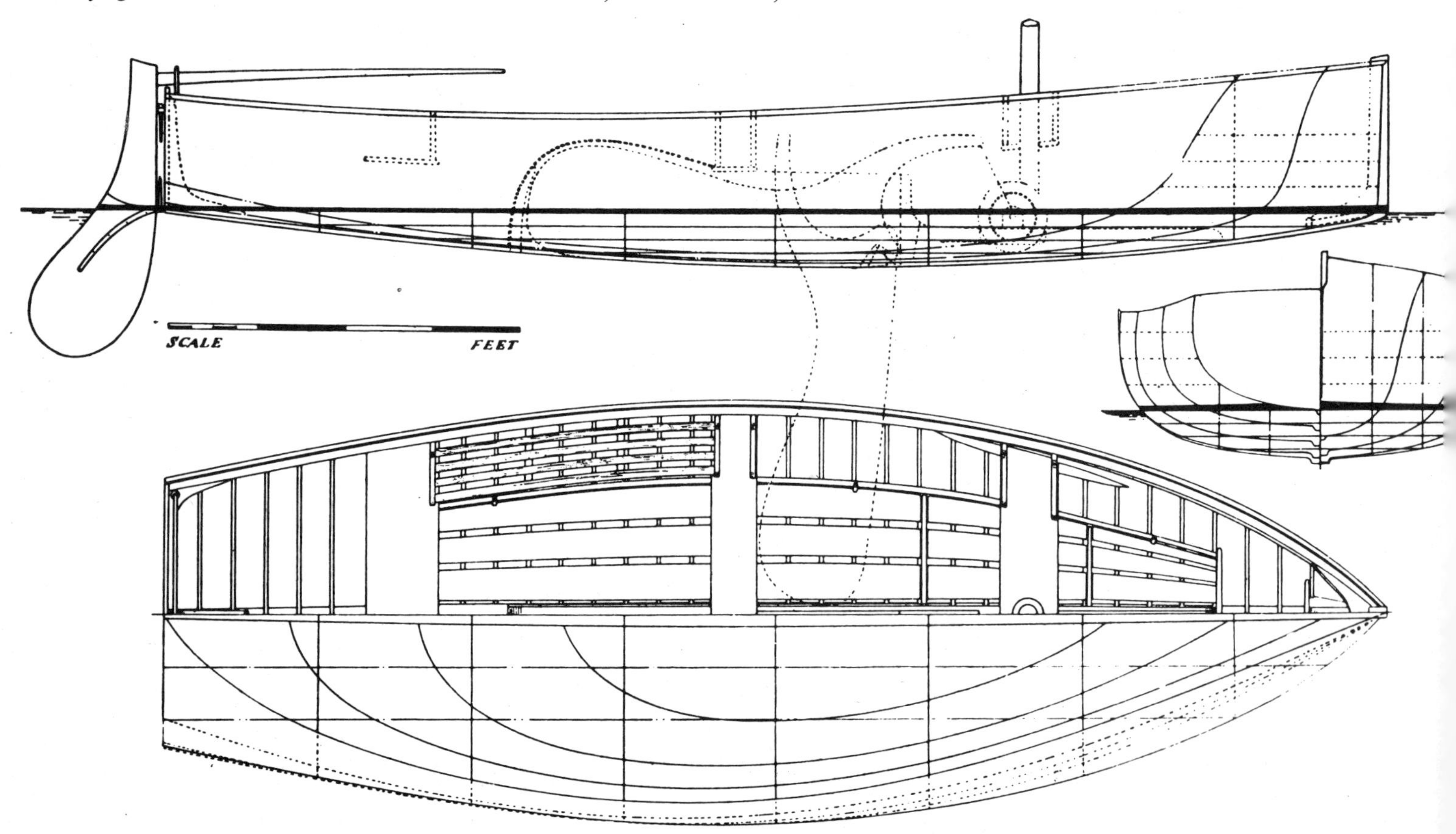

Overleaf

Morning Cloud II, built for Edward Heath in 1971. Painting by Deryck Foster, 1972.

A number of Snipe fleets grew up in Britain, but the line of future British dinghy development was most clearly foreshadowed during the 1930s in the 12-ft Nationals. When the *Yachting World* sponsored the Uffa-King design to the rules of the class it became so popular that boats to the design soon formed a

740

130
130

29744
MILIZ

Uffa Fox was a young man while the *Snark* type of boat was still sailing. He revolutionized the dinghy world with the *Avenger* type, which was immediately successful.

considerable proportion of the class. But the principle of variety and development in design was never stifled. Uffa Fox had become the leading designer in the International and National classes, and largely owing to him a technique was introduced into dinghy racing that now appears so much an intrinsic part of the activity that its absence can hardly be imagined. He taught the dinghy world to plane, or skim, and to do so not rarely in only strong conditions but as a matter of course when the wind was free. It was a revolutionary step in both design and sailing technique, and now an essential characteristic of the sport without which, to most of today's practitioners, it would be without spirit and verve.

Yet the fashionable Fourteens and the progressive Twelves of the 1930s were still rich with the racial characteristics of their inheritance. It was still evident that they were the heirs of the yacht's sailing tender. They were deckless and traditionally nautical in their aura. Even the Fourteens had something seaweedy about them.

Writing at the end of the 1947 season Ian Proctor said: 'If variety is the spice of life, the dinghy sailors' outlook should indeed be well-seasoned. For no sooner had we ceased to fire large rockets with warlike intent and commenced with less accuracy, but no less determination, to touch off small rockets to celebrate the blessings of peace, than several new dinghy classes, each with a formidable array of staunch supporters close astern, clamoured for our approval.'

The first of the new dinghy classes, which in the years immediately ahead were to roll out in an amazing virile procession, was again sponsored by the *Yachting World* magazine and known initially as the Y.W. 14-ft Restricted class, later the Merlin class, which after merging with another new class, the Rocket class, and gaining national status, became the National 14 ft Merlin/Rocket class.

The Merlin revealed the new concept of racing dinghy which was to become dominant in the years ahead. The sophisticated racing dinghy was represented

Left
A Mirror dinghy.

in 1946 by the popular pre-war 12-ft National and the exquisite 14-ft International classes. The evolution of dinghies such as these into dashing planing racing machines has been observed above. It was a characteristic of such craft that, heeled beyond a certain point, they filled rather than capsized; and that was the end of a race for them. The Merlin and later dinghies of the breed, with big areas of decking and curved side decks, might capsize without filling, be righted and sail on. The process of capsizing one's way round a course in bad weather did not immediately commend itself to the traditional dinghy racing attitude; it appeared to be an exaggerated and certainly unseamanlike procedure. Big areas of decking introduced a new dimension in dinghy racing, as earlier had the ability to plane. The post-war breed of racing dinghy shed the last of its racial inheritance from the yacht's tender. To some extent dinghy racing became a sport divorced from the maritime tradition.

The Merlin had originally been the concept of a small group of dinghy sailors, who had a boat designed by Jack Holt to incorporate their ideas. The Rocket class appeared later, in the autumn of 1949, at the instigation of former owners in the 12-ft National class who found their ages and their boats no longer perfectly compatible. With 14 ft of length the Rocket was considered to be a little less in need of acrobatics. Meanwhile the *Yachting World* sponsorship of the Jack Holt dinghy had given it, together with the set of rules for the 14-ft restricted class, much publicity. The most important rule, in the light of later dinghy developments, was that the boats should be very fully decked, the cockpit not exceeding 5 ft in length and 2 ft in breadth – a foot well. Subsequently the

National Restricted Merlin Rocket class, *Cisèle*. This class was formed by uniting the rules of the Yachting World Merlin class and the Rocket class.

more conventional Rocket class was merged with the Merlin class, and in 1950 the combined class, then totalling 330 boats, received national status.

The Yacht Racing Association had moved swiftly into the dinghy field on the return of peace, with the stated object of making small boat racing more easily available to youth. While the Merlins and Rockets were evolving, the Yacht Racing Association introduced two new classes of one-design dinghy the smaller of which, the 12-ft Firefly, being adopted as one of the Olympiad classes, became quickly established. There was nothing outstandingly advanced in the basic design; which, indeed, was so close to that of the now firmly established 12-ft

An early Firefly on sailing trials.

Trapeze work on a Flying Dutchman preparing for the Olympic Games of 1964.

Nationals that it was easy to regard the new class as a needless rival. This attitude was reinforced when it appeared that the Firefly was not appreciably cheaper than the already well established class.

But the Firefly was a step into the future recognized by few at the time. The construction of the boats was given to Fairey Marine, a maritime offshoot of the aircraft firm whose founder, C. R. Fairey, was the yachtsman who we have seen to have owned *Shamrock V* and issued a challenge for the America's Cup in yachts smaller than the accepted J-class cutters. The firm had evolved a method of moulding wooden hulls, under heat and pressure, which we may now appreciate to have been the last and most advanced technique in wooden boatbuilding, now approaching its last gasp. In 1946 the Firefly first appeared in competitive sailing; in the 1947 season the new dinghy was establishing itself, though under a few scattered clouds of doubt. The moulding of hulls, to spread so quickly during the 1960s from small to large craft and become the conventional method of construction in yachting, was strange and nearly *outré* in 1946, though the moulding was yet in loved, traditional timber. Locked in the minds of a few chemists on terms with the long molecule were the techniques that, twenty years later, were to lead to the great majority of the yachting fleet, large and small, being moulded in a mixture of glass fibre and resin, produced in factories far from the feel of salt. In 1946 there were people who would criticize the Firefly as being suitable for those who liked their dinghies to be cooked like waffles rather than built like boats. The shipwright, laying planks of carefully selected timber to create craft large or small, was still cherished reality in 1946.

The Royal Yachting Association's 15-ft Swordfish dinghy failed to gain much support, but by the mid-fifties the Association's other dinghy, the 12-ft Firefly, urged on, we have seen, by the fact of being an Olympiad class, could rival, in its strength of about 1,500 boats, the 12-ft National class, now nearing its twentieth year. But it was the latter, together with the younger Merlin/Rocket class, which kept alive the creative and experimental spirit in the dinghy world. In these two development classes there was great variety of design and ever-lively search for improvement.

This was the work of devotees. To the spectator it was the explosive growth of the dinghy world that attracted attention. In 1939 racing dinghies had still formed a tiny part only of the sailing fleet. By 1955 they had become an immense feature. The proliferation of one-design dinghy classes was being criticized. There were becoming too many of them, not differentiated enough, it was being said, to justify the multiplicity which every year became greater. A dinghy class racing in an estuary had been a refreshing sight not dimmed by undue familiarity in 1939. As the 1950s proceeded the estuaries, the lakes, the reservoirs were carrying their load of dinghies, even to suffocation in the more favoured areas. The problems involved in starting and controlling the larger classes was having to receive attention. The term 'dinghy parks' was beginning to be used, an expression allusive in a way not everyone liked.

But in a world in which dinghies were becoming numbered in millions, the multitude of classes was the only alternative to a most deadening national and international uniformity. In the days, now hard to recall, when the sailing and racing of small centreboard boats had been the uncommon taste of mainly coastal dwellers, each locality tended to have its own class, the product of a local builder. Classes of a dozen boats would be considered thriving, and there were many of them. The numerous classes of dinghy that appeared during the 1950s and 1960s were not out of proportion to the total number of dinghies that had come into existence.

Amongst this swarming fleet, the most striking feature during the fifteen years following the war was the growing proportion of hard chine, or vee-bottom, boats built from salt water resisting plywood. The sharpie hull had long before this become common in the U.S.A. The famous Star and Snipe classes were but the most famous examples in yachting of common type of small craft. Not until the invention of waterproof plywood during the war, however, could full advantage be taken of the vee-bottom form; to build this in narrow planks of natural wood produced a heavy hull which, allied to the disadvantages of the angular shape, produced a boat with a lower performance than the round-bottom type. The Snipes and Stars and others throve despite this fact.

The amateur 'build-it-yourself' principle in boatbuilding spread from the U.S.A. to Britain and Europe, and in marine plywood was found the ideal material for the technique. Using the chine hull form, kits of parts might be produced for home assembly; while the resulting boats, with their panels of thin yet strong plywood and relatively little framing, could be produced light enough for good sailing performance. Though marine plywood was a new material to the boatbuilding world, the technique of assembling boats from kits of parts was an old practice. A few centuries before the first racing dinghies had appeared, the Shetlanders had been building their Scandinavian types of boat from sets of planks, frames and shaped sterns and keels, sent from Norway.

One class of boat constructed on this principle was within a few years to become the biggest dinghy class in Britain. It accorded with the spirit of the times that the class should have been designed for young people, and appropriately named the Cadet. Sponsorship was by the *Yachting World* and the class was under the far-seeing management of the editor, Group Captain E. F. Haylock. He wrote at the time of the little boat, which was 10 ft 6¾ in in length: '. . . I felt that if the Cadet was to have a strong appeal she should also be a cheap boat.

Opposite

Bambino, of the International Cadet class which is the yet most successful racing machine for training the young.

Below

Australian sailing with big crew to match the sail area. This is an Unrestricted 18 ft class racing in Auckland harbour.

I therefore asked Jack Holt to design me a boat which would be easy for youngsters to build themselves in school carpentry classes, for their fathers to build, which would lend itself to factory production as sets of parts for assembly, or which could alternatively be built by a boatyard at a reasonable price.'

The Cadet was unique in being a young peoples' racing boat, the means of training a new generation of racing helmsmen; a hard chine pram that might be capsized without filling and be driven by an efficiently rigged sloop sail plan which included a spinnaker; a sophisticated rig but to scale with those who would handle it. The success of the conception was quickly proved.

The dinghy world of the 1970s is the yet most extreme manifestation of trends just faintly perceptible in the 1930s. It is the furthest limit to which the 'one-design' principle has yet been carried. It has made that principle acceptable in racing as it has never been in the past. The story of one-design racing is rich in class rules drawn up with painstaking care and then earnestly evaded through an enthusiasm for improvement. Hulls moulded in plastic have precluded such enthusiasm and produced a uniformity hitherto unattainable. Yet still in mass-produced dinghy classes of seemingly identical boats, in the sails and rigging arrangements, individuality persists – the individuality that a tennis player finds in his racquet.

16 Worlds of their own

All the singlehanded ocean voyages that had been made since Joshua Slocum had arrived home at Newport, Rhode Island, in 1898, after a circumnavigation of 46,000 miles, were to be remarkably surpassed in a little more than the decade following 1960. A succession of such achievements in the brief period, and the avidity with which a mainly shoregoing public followed them, may have been due partly to sociological factors beyond this study. The impersonalization of life – and death – on shore, brought a yearning, which great numbers were happy to satisfy vicariously, for the life and dangers of men alone on oceans. Paradoxically, the attention that millions were enabled to pay to these solitary exploits was due to the very conditions of the modern world against which the exploits themselves were in revolt. Newspapers with mass circulations and commanding lavish radio communications brought the solitary men in little ships on great oceans into every living-room.

The name of one man is pre-eminently associated with this contemporary development. He is Lieutenant Colonel H. G. Hasler, the retired officer of the Royal Marines who first became known to the general public as leader of the 'Cockleshell Heroes' in World War II, who since the war dedicated an acute and unconventional mind to evolving other types of cockleshell.

The idea of a singlehanded transatlantic race was his, and sprang from his dissatisfaction with the rule-controlled, large-crewed ocean-racing yachts, on the development of which the chief talents and much of the money in yachting since 1945 has been concentrated. Hasler believed that somewhere there was something better than the Bermudian rig that dominated yachting, the outcome as we have seen, of more than half a century's costly experiment. The ocean-racing yacht, in his eyes, was not the best type of sailing craft to influence the true ocean-cruising yacht, because of its multiplicity of sails, the large crew required, the high degree of skill on a wet and sloping foredeck needed to handle it, and the highly engineered rig depending on numerous highly stressed little metal parts, the failure of one of which might bring disaster.

Hasler saw the need for some form of competitive event free from the rules that force ocean-racing yachts into a conformity with an established norm that allows little real variation, which might test the new ideas essential to force a change. So came the idea for a singlehanded transatlantic race. It evolved slowly while Hasler experimented on his Scandinavian Folkboat *Jester* with various rigs and self-steering gears. The idea gained the support of Francis Chichester (later Sir) who had first become prominent in the later 1920s and 1930s for several long distance solo flights, amongst them one to Australia in 1929, across the Tasman Sea in 1931, from New Zealand to Japan in the same year A brilliant professional navigator, Chichester had little experience of sailing until the 1950s, and when he joined Hasler and four others for the first singlehanded transatlantic race in 1960, he was no experienced yachtsman. The race was sponsored by the London Sunday newspaper *Observer*, and organized by the Royal

Western Yacht Club of England in Plymouth. Criticism that might have been levelled at the possibly reckless character of the race was modified by the reputation of the club, which had handled the first Fastnet race in 1925; which had also been considered possibly reckless!

No rules governed the size or type of the boats that might enter; their character was controlled by the physical fact that a man alone had to handle them on an ocean passage the rhumb line distance of which was 2,810 miles, across an ocean area where conditions even in summer could be severe. The five boats that sailed were small craft, one of the smallest being Hasler's *Jester*. They were

Jester, with a modern adaption of the Chinese-lug rig and topsides designed for singlehanded ocean sailing, on a Folkboat hull. With H. G. Hasler, and her present owner M. W. Richey, she has competed in four singlehanded Transatlantic races and completed other long ocean passages. The rig, intended to be more suitable for such work than the conventional ocean-racing rig, has gained wide interest but no considerable number of followers.

The trimaran *Pen Duick IV*, 65 ft in length overall, was entered by Eric Tabarly for the Singlehanded Transatlantic race in 1968. She was forced to retire through gear failures, but she won the race in 1972 sailed by Tabarly's countryman Alain Colas. In this year four of the first six boats across the finishing-line were multi-hull craft.

Overleaf
Vendredi 13, a freak begot by the Singlehanded Transatlantic race in 1972.

representative of the lower sized and not inherently fast cruising yachts. The winner was Francis Chichester's *Gipsy Moth III* (it was the name all his aircraft had been given) in forty days, the largest of the five, 40 ft in length overall, and the nearest approach amongst the small fleet to the offshore racing yacht – a moderate type of cruiser-racer, in fact. Hasler, in the tiny *Jester* was second, taking forty-eight days.

The race lit the fuse to a series of increasingly ambitious singlehanded exploits in ocean voyaging and racing which even as late as 1960 would have appeared disturbing to conventional maritime attitudes. A singlehanded transatlantic race

has been held every fourth year since 1960, some details of which appear in the table:

Date	Number of starters	Single hull	Multi-hull	Length of winner	Time in days
1960	5	5	—	40 ft	40
1964	15	12	3	45 ft	27
1968	35	22	13	57 ft	26
1972	56	48	8	67 ft	20

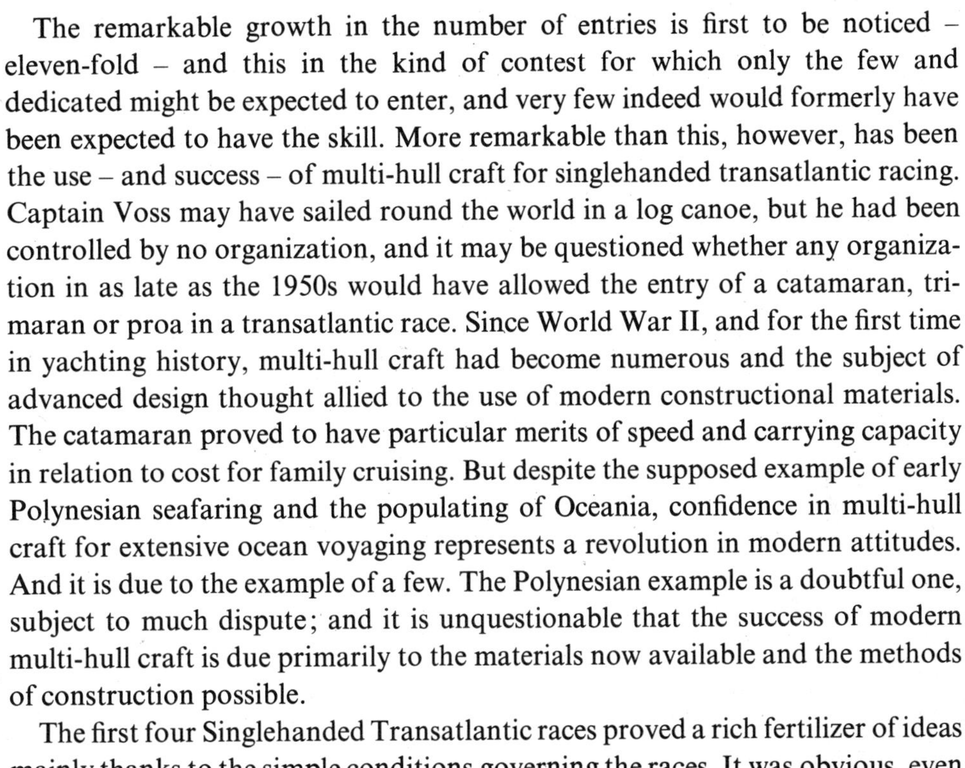

The remarkable growth in the number of entries is first to be noticed – eleven-fold – and this in the kind of contest for which only the few and dedicated might be expected to enter, and very few indeed would formerly have been expected to have the skill. More remarkable than this, however, has been the use – and success – of multi-hull craft for singlehanded transatlantic racing. Captain Voss may have sailed round the world in a log canoe, but he had been controlled by no organization, and it may be questioned whether any organization in as late as the 1950s would have allowed the entry of a catamaran, trimaran or proa in a transatlantic race. Since World War II, and for the first time in yachting history, multi-hull craft had become numerous and the subject of advanced design thought allied to the use of modern constructional materials. The catamaran proved to have particular merits of speed and carrying capacity in relation to cost for family cruising. But despite the supposed example of early Polynesian seafaring and the populating of Oceania, confidence in multi-hull craft for extensive ocean voyaging represents a revolution in modern attitudes. And it is due to the example of a few. The Polynesian example is a doubtful one, subject to much dispute; and it is unquestionable that the success of modern multi-hull craft is due primarily to the materials now available and the methods of construction possible.

The first four Singlehanded Transatlantic races proved a rich fertilizer of ideas mainly thanks to the simple conditions governing the races. It was obvious, even at the time of the first race, that the optimum boat, if of single hull type, would carry the maximum sail area that one man could handle set above the longest, lightest hull that the sail was capable of driving. It was not foreseen that widespread interest in the race and the readiness of commercial and other sponsors to finance craft built specially for the purpose, would lead, as soon as 1972, to the logically extreme type of boat becoming a reality.

A step towards it was made in the winner of the 1964 race, Eric Tabarly's *Pen Duick II*, a yacht built for conventional ocean racing but of long light form. By 1968 the pressure of competition was beginning to produce advanced boats designed for the race, and of types following lines of development that could hardly have been those expected or desired by Colonel Hasler when he proposed the race. In this year the fine entry of thirty-five included no less than twenty-two multi-hull craft. Amongst them was an astonishing 67-ft trimaran ketch in aluminium, entered by Eric Tabarly. Her displacement was a mere $6\frac{1}{2}$ tons. This craft was rushed, far from ready, to the starting-line; and though Tabarly believed her to be the right craft for the race, he would have preferred to have sailed the better prepared *Pen Duick III*. But there was only one self-steering gear available for the two boats, and Tabarly's selection fell on the trimaran despite the fact that he records 'telling myself I should be very lucky to reach the end of the race'. Many of those who saw the craft at Plymouth before the start felt likewise. *Pen Duick IV* did not get far. Suffering a collision and a number of breakdowns she returned first to Plymouth, then to Newlyn; but Tabarly was

forced to retire after a third start. His faith in this remarkable vessel was to be rewarded, however, in 1972 when she won the transatlantic race.

Also in the race in 1972 was the 40 ft proa schooner *Cheers* sailed by Tom Follett (U.S.A.). Of this craft the author wrote in the London *Observer* at the time: 'The proa is one step further than the catamaran and trimaran towards the utterly alien in western concepts of seagoing sailing craft. What can successors of the Vikings and Drake think about a marine vehicle having no definite bow, stern, port or starboard, but which goes with first one end leading then the other; and in which everything, including the rig (which is placed on one of the two hulls, not on the centreline) together with the mind of the skipper has to go into reverse every so often?' The committee was in doubts about accepting *Cheers* for the race, especially after hearing that she had capsized in the Caribbean. 'We are still of the opinion that to race along at 25 knots in between periodically capsizing is not a proper way to cross the Atlantic,' stated the secretary of the Royal Western Yacht Club. But *Cheers*, somewhat modified, was allowed to enter, and she delighted her supporters by gaining third place.

Despite the number of multi-hull craft, the winner in 1968 was a normal yacht, the 57-ft ketch *Sir Thomas Lipton* sailed by Geoffrey Williams. A long, light fin-keel yacht, derived from the faster type of model racing yacht, she was almost everything that the opinion of a generation earlier would have considered unsuitable for singlehanded sailing; but by the standards of the entries for the race in 1972 she was of an elegant and relatively traditional type. Her designer, Robert Clark, also produced Sir Francis Chichester's *Gipsy Moth III* and *Gipsy Moth V*, and Chay Blyth's *British Steel* (see below).

In the 1972 race the number of entries was more than doubled, though the number of multi-hulls fell by forty per cent. The trimaran *Pen Duick IV* was amongst them, sailed by Eric Tabarly's countryman Alain Colas. If the observers at Plymouth before the start of the 1968 race had viewed the multi-hull section of the fleet with some alarm, there was more confidence about it in 1972. This time it was another French boat of single hull form which was capable of raising alarm to match that felt about *Pen Duick IV* in 1968. Named with fitting perversity *Vendredi 13*, she was 128 ft in length and a three-masted schooner. Scarifying stories were abroad about the great yacht's unmanœuvrability and Hasler suggested that such a craft put the art of sailing back some six hundred years. The handling of so vast a ship by one man – that man looking like a fly against the huge expanse of the deck – was not a feat that could appeal to seamanlike tastes.

In the last stages of the race *Vendredi 13*, after being in the lead, was unable, with her small and inefficient sail area in relation to the size of the boat, to keep going well in light winds, and she came in some sixteen hours astern of the trimaran *Pen Duick IV*. The author wrote at the time: 'One cannot avoid being a little saddened by both the multi-hull winner and the single hull second in this year's race – saddened that such monsters (*Vendredi 13* was eighty per cent larger than the next longest single-hull yacht in the race) should have been proved by experience to be the best boats for the noble object of sailing the Atlantic singlehanded in fast time.' The 1972 race brought great credit to the multi-hulls. Thirteen per cent of the fleet was composed of this type, but four, or sixty-six per cent, of the first half-dozen arivals at Newport were trimarans – the first, third, fifth and sixth boats to finish.

During 1966–7 Chichester made his celebrated voyage round the world. One rather esoteric object of the enterprise was to make a faster time to Australia than the clipper ships that had once traded on the route. Logically, comparison between a singlehanded yacht 53 ft in length overall and a clipper more than 200 ft in length manned by an assorted but numerous crew, had no particular validity. But the self-imposed target set by Chichester of 100 days was exceeded by only a week, an amazing achievement without taking account of the navigator's age and relatively short sailing experience.

Sailing in August 1966 from Plymouth, Chichester reached Sydney in December. The return passage round Cape Horn ending at Plymouth in the following May before enthusiastic crowds completed a unique circumnavigation in striking contrast to those of such pioneers as Slocum, Voss, Pidgeon, Gerbault. None of these were in a hurry. Slocum made his circumnaviagtion in three years, three months and two days. Pidgeon took just less than four years. Gerbault's voyage lasted six years with dalliances amongst the South Sea Islands. The test of man and boat imposed by Chichester's time schedule made his voyage unique up to this time. It was an outstanding tribute to the excellence of modern yacht design and construction as well as to the indomitable spirit of a man in late middle age.

Chichester's circumnavigation was followed by that of Alec Rose (later Sir) in 1967–8. Unlike Chichester, Alec Rose aimed for no record. He intended sailing non-stop to Australia and then returning round the Horn. His yacht *Lively Lady*, of 36 ft overall and very heavy displacement, was a cruiser of good design in the inter-war manner; she epitomized the attitude of her day towards what a good seagoing yacht should be; built heavily of teak, short-ended,

Sir Francis Chichester in *Gipsy Moth IV* sailing along the New South Wales coast towards Sydney in December 1966, at the conclusion of the outward passage of his circumnavigation.

with no features conducive to speed, she was what the pre-war yachtsmen would also have regarded as being in size and type a suitable yacht for singlehanded world girdling. The venture was unsponsored, and until the yacht was approaching Australia there was little publicity. Then the enthusiasm broke which carried a quiet and modest man to the heights of popularity.

Rose experienced rigging troubles on the passage out following a succession of gales. The first failure occurred when the upper splice of the lower starboard shroud parted; next the bobstay came adrift from its stem fitting, and the port lower shroud went the way of the starboard a fortnight later. It was a chain reaction of rigging failures, which the present author discussed in an appendix to Alec Rose's book *My Lively Lady*. On the return voyage rigging trouble, probably initiated by the earlier failures, again occurred, and Rose was forced into Bluff, New Zealand. A new masthead fitting was flown out from England. The return voyage continued, followed now in the press and on the radio by thousands.

The voyages round the world of Sir Francis and Sir Alec raised a public interest in Britain which subsequent and even more exceptional voyages did not win. They were both outstanding feats of singlehanded seamanship; both triumphs of individual endeavour in a world where mass activities increasingly predominate. The crowds that greeted them, the front pages of the national newspapers and the television screens which showed the rest of the world the cheering crowds and the men themselves, grey with weariness as they landed from their lone struggles, were also demonstrations of the power that modern organized communications have to rouse the imagination of the mass mind to a frenzy of hero worship. Chichester received the accolade of knighthood on his return to England, while numerous shop windows in London displayed advertisements for goods under his photograph and recommendation. A year later Rose received his knighthood, and Portsmouth, the home of Britain's declining seapower, decked its streets as for the welcome of a conqueror.

Crowds may roar to see the bull slain, but every Everest will find its conqueror with or without the encouragement of masses. The world has been circumnavigated singlehanded under sail in quicker time than ever before, but it had not yet been girdled without a stop. This was the next Everest, neither more nor less sensible as an object than reaching the mountain's summit.

In 1968 the *Sunday Times* (London) offered an award to the first person to sail singlehanded and non-stop round the world. Some people questioned whether the sailing yacht had yet reached a state of development fit for such a challenge. But on 22 April 1969 Robin Knox-Johnston returned to Falmouth, the only entrant to complete the course in this endurance test of sailing. The 32-ft ketch *Suhaili* in which he achieved this was essentially a simple yacht, rather old fashioned in type, but beautifully balanced; which enabled her to be handled efficiently by her one crew after her self-steering gear had broken down. Robin Knox-Johnston was a professional seaman in the British merchant service; he held a master's certificate and had been first officer of the 14,500-ton British–India liner *Kenya*. His modest, quiet achievement makes one of the greatest epics of singlehanded seamanship.

After this there was, in singlehanded circumnavigations, only one more peak to conquer. In the year following Robin Knox-Johnston's return the British Steel Corporation announced that it was sponsoring a lone yachtsman's attempt to sail non-stop round the world from east to west.

It might seem perverse to want to sail alone round the world from east to west and hence, during long periods of the voyage, to be fighting against the wind

Opposite

Alec Rose in *Lively Lady* returns to the welcome of Portsmouth after his circumnavigation, 4 July 1968. Painting by Deryck Foster, 1968.

and the ocean currents in the worst of all oceanic areas. This, however, was the intention of Chay Blyth, sailing a yacht designed and built for the purpose and as advanced a sailing vessel as any yet produced for a singlehanded seaman. This fact, and some others, were able to remove the project from the category of a stunt and place it firmly in the same class as, again, the conquest of Everest.

Gipsy Moth IV in which Sir Francis Chichester made his circumnavigation in 1966. Painting by Montague Dawson. (Owned by Lady Chichester; photo by courtesy of Frost and Reed Ltd, London.)

The commercial sailing vessels, with their limited abilities to make way against foul winds, depended for parts of their voyages on the blessing of the steady Trade Winds extending some thirty degrees of latitude on either side of the equator and blowing north-east and south-east respectively in the northern and southern hemispheres. But in the great Southern Sea below the Horse Latitudes, into which they were driven by the lack of Suez and Panama canals, they were forced to round the world in the immense ocean spaces where the prevailing wind is perpetually westerly – the Roaring Forties – which sets up the West Wind Drift, or east-going current. Such ships, square rigged and heavily loaded, could not have confronted both foul wind and current. From northern waters they went east-about round the Cape of Good Hope to Australia; thence home round Cape Horn before the Westerlies and current of the Southern Ocean. We may indeed wonder how the ocean shipping of the world could have been operated with square rigged ships in the absence of the great belts of steady, if often furious, winds, of which they might take such advantage as they could. All but one of the singlehanded circumnavigators have sailed east-about; and the exception, the pioneer of them all Joshua Slocum, took the bite out of the passage by maintaining a course well to the north through the islands of Oceania and to the north of Australia.

A boat of exceptional qualities was required. Had it not been for the fact that the yacht, *British Steel*, possessed them, the venture could only have appeared extremely ill-advised. She was a large craft in terms of modern yachts, 59 ft long overall and fairly heavy displacement, ketch rigged, with a steel hull and light alloy masts. The hull was formed of mild steel plates $\frac{1}{4}$ in and $\frac{3}{16}$ in in thickness laid on steel angle-bar frames. The bulkheads were of $\frac{1}{8}$ in mild steel, and the upperworks of non-magnetic stainless steel. It is a hull of immense strength for the size of boat; in this respect, it may safely be said, far in advance of any former singlehanded yacht. She sailed from Hamble, England, on 18 October 1970 on a course direct to the Horn, through the north-east and south-east Trades and down the coast of South America, every mile bringing her nearer to the foul winds and immense seas of the Southern Ocean.

The speaking records made by Chay Blyth during his voyage – many yards of intimate film revealing a man alone facing some of the world's most hellish physical enormities – form a unique contribution to the history of seamanship; also an amazing tribute to the human spirit. He had to suffer the extremes of cold aggravated by that perpetual stunning motion and noise which give ocean seaways the power to drive men mad. With everything conspiring to produce the ultimate of lassitude, the sense that it all would never end, he had to work and keep on working – dangerous work. That Chay Blyth overcame the temptation – ever-present in such enticing colours – to steer north into quiet seas under kind skies, is the human triumph of the voyage.

'I'll never do it again, not even with a crew,' he said when he regained the Hamble river 292 days after leaving it.

In 1971 Sir Francis Chichester set out to establish another self-imposed single-handed ocean sailing record – to cover 4,000 miles at an average speed of 200

Above

Suhaili, in which Robin Knox-Johnston – seated on the cabin top – sailed singlehanded and non-stop round the world in 313 days at sea during 1968–9.

Above right

Large for a singlehander and with a steel hull of great strength, the 69 ft overall *British Steel* was sailed singlehanded and non-stop round the world from east to west by Chay Blyth.

Overleaf

At the start of the Whitbread sponsored Round-the-World race in September 1973. To the left is the three-masted *Vendredi 13* (see p. 266), not a competitor, which was second in the 1972 Singlehanded Transatlantic race.

miles a day. The numbers had no particular significance other than their roundness; but it was a soaring object as it entailed maintaining an *average* speed of 8.35 knots for some two-thirds of a month, with appreciably higher speeds being held for long periods; and this is a boat handled by one man. No fully manned conventional ocean-racing yacht has, it is believed, maintained such an average speed even for a week. There was the difference, however, that for his speed burst Sir Francis chose a course giving the likelihood of strong fair winds for a majority of the time, whereas ocean-racing courses are designed to provide a high proportion of windward work.

Down-wind courses are in the tradition of ocean-going sail, which depended, we have seen, on making use of the belts of the Trade winds. The course selected by Chichester was from Bissau in Portuguese Guinea to San Juan del Norte (Greytown) some 120 miles north of Panama. The speed of ships being partly a function of size, the speed required to meet what Chichester called 'The Romantic Challenge' had of necessity to be large for one man to handle. He selected a yacht with the same hull as that of Geoffrey Williams' *Sir Thomas Lipton*, in which he had won the Singlehanded Transatlantic race in 1968, but with a rig modified to suit down-wind conditions. The new yacht, *Gipsy Moth V*, had a length overall of 57 ft, was of light displacement and carried an unusual version of the ketch rig, with a mizzen staysail but no mainsail, and several running headsail each of greater area than any other of the individual sails.

As the present author wrote at the time (republished as an appendix to Sir Francis Chichester's book *The Romantic Challenge*) 'The crucial technical factor that must make Chichester's objective so difficult to achieve is due to the size of the boat in relation to the required average speed . . . to maintain this average [the speed required for an average of 200 miles a day] the yacht must spend part of the time sailing at about 9.5 knots . . . Sir Francis, alone and with a 57-ft two-masted yacht on his hands, must for some periods flog her up to

VENDREDI
13

Opposite
Gypsy Moth V.

Overleaf
Regatta on the Starnberg lake, 1882, or, as we were.

and beyond this speed, near the hydrodynamic limit of performance, and hold her there . . . It is the fact that the yacht must be held for such periods in the extreme upper part of her speed range that gives the operation an air of impossibility, and makes one wonder whether anyone but Sir Francis would have the confidence to attempt it . . . The voyage may be no less than 4,000 miles, but he will not be able to ease off for even a few seconds. The Trade winds will have to behave immaculately. Calms or light winds will cripple the venture.'

The voyage extended from 11.30 hrs on 12 January to 19.25 hrs on 3 February. The target speed was not reached; but the performance was impressive enough. It was closer to the target than most judges anticipated. The time for the 4,000 miles run was 22⅓ days, giving an average of 179.1 miles a day. In the course of the whirlwind rush two spinnaker poles were broken, and for three days and eight hours of the last five days, during which the average of 200 miles a day *was* maintained, the two poles were lashed together. The actual average speed was fractionally less than 7.5 knots.

An historically valuable feature of the romantic challenge was the accurate record of speeds under sail it was able to provide. Speed claims under sail are notably unreliable; those made for the clipper ships and still widely accepted will not stand a careful navigational analysis; those for yachts are liable to have even less reality. Sir Francis Chichester, as an expert navigator with a specialized interest in the speed problem, was able to provide in his log of this voyage valuable data on speed under sail which henceforward, whenever the subject is under discussion, no one will be wise to neglect. He has provided a yardstick of the possible though, in fact, rarely attainable.

Shortly after his return Chichester wrote to the author: 'In a storm her speed was a great disadvantage! I could not slow her down. Running under bare poles, I once noted the speedometer reading 12½ knots. When I tried to slow her down she would not point into the wind with the helm hard a'lee. I reckon she might have sailed into the wind at eight knots if I had trimmed for it, but this would have been suicide.'

At the time of his death Sir Francis was planning a further romantic challenge to reach the average speed of 200 miles a day over 4,000 miles. 'I became enchanted by the neatness, the rightness of the figures . . . They seemed as attractive to the solo sailor as the 10-second hundred yards and the 4-minute mile had been to athletes before those targets were achieved.'

At the beginning of this book the author tried to summarize the many natures of yachting. The rest of the book has been an attempted demonstration of this. But a volume could be written for each chapter; for the post-1939 period alone a volume of this size would not be inappropriate. The illustrations and their captions will help to fill out what must have been in some degree a selective treatment, as they will certainly impress vividly on readers the multiplicity and beauty there has been in the world of yachting.

BAVARIA

Select bibliography

A yachting library can now be immense. The following works have been selected as being particularly valuable to the historian

Archer, James, *Colin Archer – A Memoir*, Gloucester, U.K. 1949

Atkins, J. B., *Further Memorials of the Royal Yacht Squadron 1901–38*, London, 1939

Beken & Uffa Fox, *The Beauty of Sail*, London, 1938

Beken, Frank & Keith, *The Beauty of Sail*, London, Toronto, Wellington, Sydney, 1964

Burgess, Edward, *American and English Yachts*, New York, 1887

Chamier & Beken, *Beken of Cowes 1897–1914*, London, 1966

Chamier & Beken, *Beken of Cowes (2) 1919–39*, London, 1969

Chapelle, Howard, *The History of American Ships*, New York, 1936

Clark, Arthur, *The History of Yachting 1600–1815*, New York, 1904

Coles & Proctor (Edited), *The Yachtsman's Annual 1949–50*, London, 1950

Coles & Proctor (Edited), *The Yachtsman's Annual 1950–51*, London, 1951

Crane, Clinton, *Yachting Memories*, New York and Toronto, 1952

Folkard, H. C., *The Sailing Boat*, London, 1906

Fox, Uffa, *Sailing, Seamanship and Yacht Construction*, London, 1934

Fox, Uffa, *Uffa Fox's Second Book*, London, 1935

Fox, Uffa, *Sail and Power*, London, 1936

Fox, Uffa, *Racing, Cruising and Design*, London, 1937

Fox, Uffa, *Thoughts on Yachts and Yachting*, London, 1938

Gavin, C. M., *Royal Yachts*, London, 1932

Gerbault, Alain, *The Fight of the Firecrest*, Paris, 1930

Guest & Bolton, *Memorials of the Royal Yacht Squadron 1815–1900*, London, 1903

Haylock, Phillips-Birt & Hayman (Edited at various periods), *Yachting World Annual*, London, annually 1951–67

Heckstall-Smith, Anthony, *Sacred Cowes*, London, (revised) 1965

Heckstall-Smith, B., *Yachts & Yachting in Contemporary Art*, London, 1925

Heckstall-Smith, B., *Britannia and Her Contemporaries*, London, 1929

Hiscock, Eric, *The Yachting Year 1946–7*, London, 1947

Hiscock, Eric, *The Yachting Year 1947–8*, London, 1948

Hofman, Erik, *The Steam Yachts*, Lymington, 1970

Hoyt, Sherman, *Memoirs*, New York and Toronto, 1950

Hughes, John Scott, *Famous Yachts*, London, 1929

Hughes, John Scott, *Sailing Through Life*, London, 1947

Irving, John, *The King's Britannia*, London

Irving, John, *British Yachts and Yachtsmen*, London, 1907

Kemp, Dixon, *Yacht and Boat Sailing and Yacht Architecture* (11th Ed.), London, 1913

Knight, E. F., *The Falcon on the Baltic*, London

Kunhardt, C. P., *Steam Yachts and Launches*, New York, 1887

Kunhardt, C. P., *Small Yachts*, New York and London, 1885

Lambert, Gerard B., *Yankee in England*, New York and London, 1937

Lawson & Thompson, *Lawson's History of the America's Cup*, Boston, 1902

Leach, George & Others, *Badminton Library: Yachting* (2 vols), London, 1894

Loomis, Alfred, *Ocean Racing*, New York, 1936

Martin, E. G., *Deep Water Cruising*, London, 1928

Mead, C. J. H., *History of the Royal Cornwall Yacht Club 1871–1949*, London, 1951

Parkinson, John, *Nowhere Is Too Far*, New York, 1960

Philips-Birt, Douglas, *Fore & Aft Sailing Craft*, London, 1962

Phillips-Birt, Douglas, *British Ocean Racing*, London and New York, 1960

Phillips-Birt, Douglas, *An Eye for a Yacht*, London, 1955

Poor, Charles Lane, *Men Against the Rule*, New York, 1937

Rayner & Wykes, *The Great Yacht Race*, London, 1966

Rosenfeld, Morris, *The Story of American Yachting*, New York, 1958

Schoettle, Edwin J., *Sailing Craft*, New York, 1945

Slocum, Joshua, *Sailing Alone Round the World*, New York, 1901

Stephens, W. P., *American Yachting*, London and New York, 1904

Stone, Herbert L., *The America's Cup Races*, New York, 1930

Vanderbilt, Harold S., *Enterprise*, New York and London, 1931

Vanderbilt, Harold S., *On the Wind's Highway*, New York and London, 1931

Vanderdecken, *Yarns for Green Hands*, London, 1862

Voss, John, *The Venturesom Voyages of Captain Voss*, London

British Journals

The Yachtsman (founded 1891)
Yachting Monthly (founded 1906)
Yachting World
The Field (some issues)
Country Life (some issues)
The Motor Boat

U.S.A. Journals

Rudder
Yachting
Pacific Motorboat
Motor Boating

Abbreviations

Beken of Cowes, Isle of Wight	Beken
Morris Rosenfeld and Sons, New York	Rosenfeld
The National Maritime Museum, London	Maritime Museum
The Science Museum, London	Science Museum
The British Museum, London	British Museum
The Peabody Museum, Salem	Peabody Museum

Illustration acknowledgments

The photographs taken at the National Maritime Museum, The Royal Thames Yacht Club, The Royal Ocean Racing Club and the Malcolm Henderson gallery are by Derrick Witty; those taken at the New York Yacht Club are by Stanley Rosenfeld

Jacket illustration: Courtesy of N. R. Omell, London

Endpapers illustration: Mary Evans Picture Library

Page
2 Mary Evans Picture Library
8 From an unknown collection. Photo Michael Robinson
11 Science Museum
14 Courtesy of Rupert Preston Marine Paintings, London
17 Reproduced by gracious permission of Her Majesty the Queen
18 Maritime Museum
20 British Museum
21 From *Badminton Library: Yachting*, London, 1901 edition
22 Royal Thames Yacht Club, London
23 Science Museum
25 From *Memorials of the Royal Yacht Squadron 1815–1900*, by Guest and Bolton, London, 1903
26 Maritime Museum
29 Maritime Museum
32 Museum of the City of New York
35 and 36 Peabody Museum. Photo Mark W. Sexton
38 Maritime Museum
39 Stevens Institute of Technology, New Jersey
40 From *Lawson's History of the America's Cup*, by Lawson and Thompson, Boston, 1902
43 Science Museum
47 Mary Evans Picture Library
50 Maritime Museum
51 Science Museum
53 *Above* Collection of G. E. J. Robertson
Below Shelburne Museum, Vermont
54 Maritime Museum
56 J. Clarence Davies Collection, Museum of the City of New York
57 Beken
58 Courtesy of N. R. Omell, London
59 Science Museum
61 Peabody Museum
66/67 Beken
69 Peabody Museum
71 Maritime Museum
72 Lloyds Register of Shipping, London
74 Beken
75 Peabody Museum
79 Peabody Museum
81 The Tate Gallery, London
82/83 Royal Dart Yacht Club, Devon
84 Penobscott Marine Museum, Maine
86 Beken
88 Peabody Museum
89 Royal Thames Yacht Club, London
93 Rosenfeld
95 Beken
98 Mary Evans Picture Library
100 From *British Yachts and Yachtsmen*, London, 1907
101 Beken
103 Rosenfeld
105 Beken
107 From *British Yachts and Yachtsmen*, London, 1907
109 Maritime Museum
110 *Both* New York Yacht Club
112 Courtesy of N. R. Omell, London
116/117 The Old Print Shop, New York
120 Beken
121 Beken
123 Royal St Lawrence Yacht Club
124 Beken
125 *Above* Beken
Below Rosenfeld
127 Philadelphia Museum of Art. Given by Mrs Thomas Eakins and Miss Mary A. Williams. Photo Alfred J. Wyatt
128 Peabody Museum. Photo Mark W. Sexton

131 From *Down Channel* by R. T. McMullen, 1893

132 Photo Winfield Scott Clime, 1907. Courtesy of Walter Teller, author of *Joshua Slocum*, 1970

133 Courtesy of Walter Teller, author of *Joshua Slocum*, 1970

136 J. Dutilh, Rotterdam

138 Beken

139 Mary Evans Picture Library

140 John Topham, Sidcup

143 Norsk Folkemuseum, Oslo

145 Beken

147 Rosenfeld

149 Rosenfeld

150 Beken

151 *Top and bottom* Beken
Centre Rosenfeld

153 The Ella Gallup Sumner and Mary Catlin Sumner Collection, Wadsworth Atheneum, Hartford

154/155 George Rainbird Ltd. Photo Derrick Witty

156 Maritime Museum

158/159 Beken

161 Beken

163 Beken

166 Beken

168 Rosenfeld

170 Rosenfeld

172/173 Beken

174 Beken

176 Courtesy of Sir Thomas Sopwith. Photo Rosenfeld

179 Rosenfeld

180/181 Rosenfeld

182 Beken

183 Rosenfeld

184 Rosenfeld

185 *Above* Beken
Below Rosenfeld

187 Rosenfeld

190 Bath Marine Museum, Maine

193 Rosenfeld

195 Beken

196 SEA Publications Inc., Costa Mesa, California

197 New York Yacht Club

198 Courtesy of Malcolm Henderson, London

200 Rosenfeld

201 Rosenfeld

203 Rosenfeld

206 Beken

207 Beken

209 Royal Ocean Racing Club, London

211 Rosenfeld

213 Rosenfeld

214 Beken

215 Royal Thames Yacht Club, London

216 New York Yacht Club

220 Beken

221 Beken

222 Rosenfeld

224 Beken

227 London Express News and Features Services

228 Beken

230/231 Royal Thames Yacht Club, London

233 Collection of Arthur E. Rheinhold

234 Collection of Desmond Mollins

236 Rosenfeld

238 Press Association Photos Ltd

240 Bermuda News Bureau

241 Beken

242 Popperfoto

244 Photo Neptune

249 Rosenfeld

251 Beken

252/253 From the Collection of the Rt. Hon. Edward Heath

254 Beken

255 Beken

256 Beken

257 Beken

258 Popperfoto

260 Auckland Star

261 Beken

263 Beken

264 Press Association Photos Ltd

266/267 Photo Chris Smith, courtesy *The Observer*

269 John Fairfax, Feature Services, Sydney

271 Collection of G. E. J. Robertson

272 Collection of Lady Chichester

274/275 London Express News and Features Services

275 British Steel

276/277 Thomson Newspapers Ltd

278 Photo Chris Smith, courtesy *The Observer*

280/281 Mary Evans Picture Library

The lines plans are reproduced from the following sources:

41 *Motor Boating*, New York, 1945

44, 62, 63, 77, 171 The following yachts' lines from 1935 *Transactions* published by The Society of Naval Architects and Marine Engineers are reproduced herein by permission of the aforementioned Society: Volante, Puritan, Clara and Gloriana from *Yacht Measurement* by W. P. Stephens; Enterprise from *The America's Cup Defenders* by C. P. Burgess

51 *An Eye for a Yacht* by D. Phillips-Birt, Faber and Faber Ltd, London, 1955

80 *American Yachting* by W. P. Stephens, Macmillan Inc., New York, 1904

91 *The King's Britannia* by John Irving, Seeley Service & Co. Ltd, London

129 *The Sailing Yacht* by J. Baader, Adlard Coles Ltd, London, 1965 (Rights: Verlag Delius Klasing & Co., Sickerland)

162 *Sail and Power* by Uffa Fox, Peter Davies Ltd, London, 1936. By kind permission of Uffa Fox Limited

170, 210, 250 *Sailing, Seamanship and Yacht Construction* by Uffa Fox, Peter Davies Ltd, London, 1934. By kind permission of Uffa Fox Limited

204 *British Ocean Racing* by D. Phillips-Birt, Adlard Coles Ltd, London, 1960 (Rights: Oxford University Press)

248 *Yachting World Annual*, 1968, ed. D. Phillips-Birt, The Butterworth Group, London. By kind permission of George Philip & Sons Limited

Index

Figures in *italics* refer to the illustrations